THE 300,000-MILE CAR

Roy Cox
Davidson Gigliotti

DELMAR

™

THOMSON LEARNING

Australia Canada Mexico Singapore Spain United Kingdom United States

THE 300,000-MILE CAR

Roy Cox
Davidson Gigliotti

DELMAR

THOMSON LEARNING

Australia Canada Mexico Singapore Spain United Kingdom United States

DELMAR
THOMSON LEARNING ™

The 300,000-Mile Car
by Roy Cox and Davidson Gigliotti

Business Unit Director:
Alar Elken

Executive Marketing Manager:
Maura Theriault

Development Editor:
Christopher Shortt

Executive Editor:
Sandy Clark

Executive Production Manager:
Mary Ellen Black

Production Editor:
Tom Stover

Acquisitions Editor:
Jack Erjavec

Production Manager
Larry Main

Marketing Coordinator
Brian McGrath

Team Assistant
Bryan Viggiani

Channel Manager
Mary Johnson

Cover Design:
Michael Egan

Cataloging-in-Publication Data

Cox, Roy, 1948-
The 300,000-mile car / Roy Cox, Davidson Gigliotti.
      p. cm.
Includes index.
ISBN 0-7668-3176-0
  1. Automobile--Maintenance and repair--Popular works. 2. Automobile ownership--Popular works. 3. Automobiles--Reliabilitiy--Popular works.
I. Gigliotti, Davidson. II. Title.

TL152 .C63 2001
629.28'72—dc21
                                    2001047172

NOTICE TO THE READER

CONTENTS

This book is dedicated to my family, Nancy, Lauren and Jordan, who shared my efforts and put up with my late nights and mood swings, and to all the automotive professionals who have shared their knowledge and skill with me over the past 37 years.

Roy Cox

My work on this book is dedicated to Elaine Summers, the wonderful person who encouraged me at every point, and whose idea it was that Roy and I should write this book together. Thank you, Elaine, for everything.

Davidson Gigliotti

ACKNOWLEDGEMENTS

All the insurance executives who held our hand through the insurance section: Dave Hurst, State Farm Insurance; Lowell Olsen, State Farm Insurance; K. Doyle, Allstate Insurance; Kitty Miller, Farmers Insurance Exchange; Phil Nuss, Nationwide Insurance Agent, Nokomis, Florida; Walter Smith, Geico; Dick Hospital, Assistant Vice-president of Underwriting, Geico; Bill Hardy, Insurance Services, AAA.

The folks at Delmar Publishing who believed in our project and started us on the road to publication: Jack Erjavic, who signed us up,

Chris Shortt, who told us we were "good authors;" Tom Stover, who kept us on track.

Pam Daniel, Editor-in-Chief, Sarasota Magazine, who assigned Davidson to the story on which he met Roy.

Matthew Edlund, M.D., a fellow author and encouraging friend who listened to Davidson's many complaints. Allen Gallup, who also listened, but with a lot less patience. Emily Harvey, who let Davidson work in her living room in Venice, Italy, for two summers and thought we'd never finish.

George Giek, who hired Roy after he moved to Florida and sent him on the assignment where he met Davidson and where the idea for this book was born.

Marie Fields, engineer at General Motors, for helping Roy to sort out the finer details of the latest technology.

Unless you live right down the block from the supermarket, the doctor's office, and the local schoolhouse, and work at home to boot, you need a car to get along. This is a fact of life.

According to the Bureau of Transportation statistics, the average cost of operating an automobile in 1997 (the latest year for which we have this statistic) was $6,723. That includes variable costs like fuel, maintenance, and tires; and fixed costs like insurance, license and registration, taxes, depreciation, and finance charges. That's a lot of money to most of us. We don't believe that that number has gone down since then. And sometimes, we're sure, you must wonder if it's worth it. It has to be, of course. We've designed a world where cars are a necessity in most places.

But owning a car, insuring it, making the payments, maintaining it, and keeping it running are not simple matters—unless you're wealthy, and even then car ownership can be complicated and frustrating. Medical emergencies aside, car emergencies can put the greatest strain on household budgets for folks with ordinary incomes, simply because the car is so essential and its repair is, therefore, urgent. If you are running two cars in your family, multiply that by two. Cars are a major cost center in American family life.

We believe that we can help you get the most out of your car, and relieve at least some of the frustration as well. This book can help you through every phase of car ownership, from deciding what car to buy or lease in the first place, to communicating effectively with the service technician when things go wrong, and redressing grievances when communication fails. Not only do we hope to save you money, but we hope to help you keep your car running for a long time and help you derive the most enjoyment from it.

Along the way we try to answer every question we can think of about a whole range of car-related issues. We also fully describe the functioning of the modern car, and include troubleshooting processes for resolving problems as they arise. And we do not shy away from getting into the complexities of the digital age as it affects cars, though we hope we have made it, to borrow a phrase from the old Popular Mechanics, "written so you can understand it." We will also give you lots of information about what to look for when shopping for such things as tires and batteries, and some things you can check yourself to help keep your car in top form.

Both of us are seriously impressed with the fact that an ordinary car, these days, costs $20,000, and an ordinary gallon of gas costs upward to $2.00. We can remember when a gallon cost a quarter, and you could buy a new Ford, or a new MGA, for that matter, for $3,000. Times change, and so does the value of money, but we both suspect that the wages of most working people have not quite caught up with the rise in prices for cars, fuel, and many other products. We live in a world where both husband and wife work hard to keep up with the bills. We both remember a time when this was the exception rather than the rule.

We realize that over 30 percent of the cars on the road today are older than twelve years, and that the average age of all American cars is nearly nine years, the highest it has been since after World War II. Some of these cars, perhaps, are the pampered pets of car hobbyists, but we know that most of them are on the road because their owners have no other choice. Our hats are off to the people who own those cars and keep them running every day. They probably spend more than the $653.00 per year that the average household pays in car maintenance per year, but they avoid the big down payment and high monthly payments that a new car, or even a good used car, will cost. For many of us, the best car is one that is all paid for.

That said, part of the focus of this book is on cars of the last ten years. The old cliché, "They don't build them like they used to," was never so true! During this period cars have undergone fundamental changes in the way they work. That period has seen the end of the carburetor, the decline of drum brakes, the primacy of front-wheel drive, the introduction of light-weight mechanical components, and the almost total adoption of digital technology as the chief regulating factor in car performance. The operating systems in today's cars use computers, and not only does this make service much more efficient, it makes the car's operation more efficient, too. So, in a sense, while cars are more expensive than they used to be, you do get a lot more for your money.

Still, there is a cost. Back before computerization, cars were relatively simple machines, and their operating principles could be fully grasped by many car owners. Ordinary people used to work on their cars years ago, changing out worn parts, doing their own routine maintenance—keeping them running smoothly. For many, it was a satisfying and money-saving hobby, and for some it was a passion.

Today, cars don't easily lend themselves to that. Working on today's cars, with their densely packed engine compartments, microprocessors, relays, and sensors, often requires special tools, lifts and hoists, computer diagnostic equipment, and very specialized training. Service is expensive, too. In some shops they wear lab coats and surgical gloves.

That doesn't mean that you can't work on your car, of course. In fact, we encourage your involvement in your car's maintenance and repair, and give you suggestions for how to go about it. Even today, many car problems are still mechanical in nature, and an aggressive approach to routine maintenance will go a long way to keeping your car on the road.

Our strategy is to help you cut down on service by helping and encouraging you to become familiar with your car, and showing you how to use that knowledge to give your car a fighting chance to make it to the 300,000-mile mark without too many costly repairs. If you follow our guidance, we guarantee you will have a more enjoyable, and more economical, relationship with your car than those who take a more passive approach to car economy.

There's another benefit, too. We think that people who love their cars, understand how they work, and are actively involved with their care, tend to be better, safer, drivers. It is our earnest desire to promote safety, sobriety, and informed car ownership on our highways. So good luck to you, and to your car, and, as the old Esso signs used to say, "Happy Motoring!"

DAVIDSON TELLS WHO WE ARE AND HOW WE MET

Roy and I met in 1995 at a Mercedes-Benz press bash in Mt. Dora, Florida. Seated at a lunch table at the beautiful Mt. Dora Inn with about eight other automobile writers, I noticed that although the conversation was mostly lively and loud, Roy spoke only rarely. But when he did, the table went silent and the veteran automotive experts paid full attention to his words. No one challenged his opinions. Roy was an expert's expert.

As the event went on, Roy and I got to know each other better and I learned that Roy was manager of the Automotive Technical Training and Research Group at AAA National Offices in Heathrow, Florida. It was only much later that I learned that Roy was one of only 234 professionals in the United States currently certified by The National Institute for Automotive Service Excellence (ASE) as a Master Automobile Technician, Master Heavy Truck Technician, Master Collision and Refinish Technician, and Advanced Performance Specialist. He was inducted into the Auto Technician's Hall of Fame by the Automotive Service Industry Association (ASIA) in 1996 as a World Class Technician.

He is also a fine technical writer, who produces the AAA towing manual, lockout manual, and numerous training programs and videos for AAA towing and road service drivers. He also conducts seminars and public demonstrations, and participates in events around the nation. Roy has been fixing cars since age sixteen, and has worked on nearly every type of vehicle. He still works on cars part-time, mostly to keep his skills up-to-date. His main job now is training other technicians, but he still calls himself a mechanic. He's a pretty good organist, too.

After a short acquaintance, I suggested that we could write this book together, and Roy readily agreed. Over the intervening years we polished our ideas and our book proposal, found a publisher, and set to work. This book is the result.

As a writer I've covered many different assignments, cars included. I was an unspectacular backyard mechanic in my New England youth, mainly interested in British sports cars. My path through life has not been as straightforward as Roy's. I started out in the arts, as a sculptor, a conceptual artist, video artist, arts administrator, and art writer. I had a more or less typical writer's and artist's life, working at various trades and having adventures. I was a journalist, an architectural woodworker, a columnist, an independent video producer, and an editorial writer. I even ran a small art museum once. I've lived most of my life in New York City and I'm going back, though I live in Florida now. I've owned numerous cars in my life and have always been impressed with how much keeping a car costs. I've also been impressed with the fact that even the nicest cars will betray you without notice if they are not maintained properly.

CHAPTER 1

CHOOSING THE RIGHT CAR (IN THE FIRST PLACE)

This book is about how to get the most out of the car you own. It also includes a pretty detailed (and, we hope, entertaining) account of how today's cars work, and what to do when things go wrong.

The first two chapters, however, are about helping you to choose and buy a car that conforms to our philosophy of car ownership. Briefly stated, that is: buy a good car, pay as little as you can for it, learn about it, take good care of it, and drive it till the wheels fall off. Do what we recommend and you will save a large sum of money during your life. To get the best results from this book, you will have to put in the time. But it is worth it. And one of the first things you will want to do is read this whole book!

Right now, either you already have a car that you are committed to keeping for a while—perhaps a long while—or you are at the point of buying a car that you expect to drive for a long time. If you already have a car you intend to keep, then you might want to head right for chapter three.

However, if you are about to buy a car, you are in a slightly more advantageous position. Advantageous, because not all cars, new or used, are created equal. Some have a greater potential for longevity than others do, just like people or pets. Part of the purpose of this opening chapter is to help you identify the car that is right for you.

Only you know how much car you can afford. We can help by bringing up some of the elements of car ownership cost, and we can explicate them pretty well. But in the final analysis, you have to decide what car to buy. We can't really do that for you—we don't recommend makes and models. But we can help you make an intelligent choice. A wise marketer once said on TV, "An informed consumer is our best customer." Sadly, that philosophy is not widely embraced by the automotive industry and the financial institutions that serve it. The fact is that the job of the auto dealer's sales force is to make as much profit as possible on each sale, and that is most easily achieved if you are uninformed.

We believe you can win the automobile game by setting up a different set of expectations for yourself than the ones the automobile industries want you to choose. As we have said before, and will no doubt say again, take good care of your car, keep it a long time, and drive it till the wheels fall off. That is the way to get the absolute most out of your car.

Make no mistake. If everyone looked at cars the way we do, our automobile companies, and the industries that serve them, would be a lot smaller. We're not worried; we know that our program, if carried to its conclusion, is not for everyone. But we believe this book is for everyone. Even if you trade your car every few years, or even more often than that, there is information here that will help you achieve a better relationship with your car, and some valuable advice to help you choose the purchase or leasing plan that is best for you. If you read this book, you will come out knowing a lot more about today's cars than you did before you started. That kind of information is always useful, no matter what your plan for your car.

So inform yourself, so that you can both make wise personal decisions and hold the profits of the dealer and the finance company to a minimum. They are both good goals! We are here to help!

As we get into the body of this book you will learn from us that cars have undergone tremendous technological changes in the last ten years, due mainly to the ever-increasing use of electronics. Not only that, but also digitalization and the Internet now influence the way we decide what car to buy and, even, as you will learn in the next chapter, the actual buying of the car itself.

Looking at the way most of us live these days, we can see why the Internet is such a great tool. Most of us are stretched extensively, both at work and in our personal lives, to do more and more activities, to "work smarter" and "think outside of the box." All of this translates into finding creative ways to carve up our days even more than they are already and pack in more so-called "productive time." Some publications we have read actually indicate that the average American citizen has a whopping three hours of discretionary time (not committed to someone or something) per week! If you look around you, it's easy to believe that the number is getting even lower! It is no wonder that we relish what little free time we have and try our best to keep some sense of control. Add an unexpected event to the week and our free time has vanished. Once more, we are looking for a way to pack even more things into our busy schedule in order to deal with the new demands.

Unfortunately, this typical scenario puts us squarely at odds with what we really must do to protect our interests as we go through the steps of choosing and acquiring a car or truck: spend time! The theme of this chapter and of chapter two could be, "above all, do not act in haste!" In order to get the best deal, obtain the best financing arrangements, and buy the right equipment, we must do our homework and do it with care.

About Today's Cars...

Over the last decade, cars have undergone radical changes, some driven by regulation and others due to the changing desires of the motoring public. The common link is more high-tech, digitalized equipment. We will discuss this more in later chapters, but one point we want to make is that the cars of today are much improved over the cars of yesteryear. When considering the suitability of most cars today for the 300,000-mile program, we find that most of today's car manufacturers, domestic and imported, make cars that will go the distance. That wasn't always the case. When we were young (and that was some time ago) a car that made it to the 100,000-mile mark was a local celebrity, especially if the engine had never been "torn down" for an overhaul or a valve job. Cars were not as well made or, as importantly, as well engineered. People didn't drive so much then—12,000 miles a year was a lot; a car could go for eight years at that rate.

So in your search for reliability, safety, good repair history, and endurance, you have a pretty wide field from which to choose. Finding the right car for you, one that you can drive in comfort, and with confidence and a measure of pride for fifteen years or so, is an ambition well within your reach. All you need are the resources and the time to do the research.

First Stop: The Internet!

If we had written this book five years ago, we probably would not have written this section. But the Internet has become such a valuable research tool that we see it as your first stop on the way to current automotive knowledge. If you are contemplating the purchase of a car, and you have access to the Web, you have at your fingertips one of the best resources for new and used car information in the world. Type a keyword or phrase such as "automobile buying" into your favorite search engine and you will see a list of Web sites that could keep you surfing for days on end. Here are just a few of the sites, with a brief description of each, that we have found valuable for prospective car buyers. We have included other useful sites in later portions of the book, as well.

http://www.autoweb.com/sitemap/default.htm

This is a very rich site that is fun to explore. Start with the site map and check out a long and well-organized series of links. You can buy a car, finance a car, lease a car, sell a car, and buy insurance, all on this site. No, they don't deliver the car next day by Airborne Express. What they do do is refer you to another Web company, in this case *CarsDirect.com*, which in turn quotes you a price for the option package you choose and sends you off to a dealership near you that is part of their program. Not bad.

The site also offers a pretty good range of research tools, maintenance advice, and many other bells and whistles. It will also direct you to...

http://www.carfax.com/

For $19.95, Carfax will run title search reports for as many vehicles as you desire within a 60-day period. A single report costs $14.95, so it's well worth the difference if you are in the throes of car shopping! Using the car's vehicle identification number (VIN) you can find out some interesting facts, including exact vehicle specifications, title history, odometer check, and an interesting assortment of possible negatives like: junk/salvage title issued, manufacturer's buy back ("Lemon Law Title") issued, flood damage title issued, damage disclosure title issued, police department accident report, etc. It's always nice to know you are not buying a stolen car! Davidson found out that his '93 Integra, which he bought used in Florida, came all the way from the state of Washington. Carfax also offers a variety of other services, including a 30-day, money-back guarantee on the car you purchase. It's a Web site worth visiting!

http://www.insure.com/auto/

This is a very complete in-depth insurance site with excellent tools for insurance research. It features an exhaustive list of questions and answers, articles, and other information. They even rate insurance companies by financial strength! Really, this is a must for prospective buyers of car insurance.

http://www.nhtsa.dot.gov/cars/

This is the site of the National Highway Traffic Safety Administration. Their main headings are: Safety Problems and Issues, Testing Results, Regulations and Standards, and Research and Development—each with a long list of links. Of major interest to car buyers is their crash test program. Type in make, model, and year, and chances are they've crash-tested it and will give you the results—very useful stuff.

http://www.carsafety.org/

This is the site of the Highway Loss Data Institute, a nonprofit group sponsored by the insurance industry to provide some of the information that insurers use to set their rates. It is also a very rich site, with good publications. Of particular interest to the car buyer is "Injury, Collision, and Theft Losses." You can download this for free and read it as a PDF file, but special software may be required to open the file. It contains invaluable safety and theft information about most car models from 1997 to 1999. They are always a year or so behind, because their information is collected from field research, not from the automotive industry. But if you are buying a used car in that date range, you will want to see this information. They also provide crash-test results, and much other information.

http://www.nadaguides.com and http://www.kbb.com/

The information contained in both the *Kelley Blue Book* and the *N.A.D.A. Official Used Car Guide* is available on-line and can be accessed without charge. Just type in the year, model, and mileage, and they will come up with the trade-in value and the retail value. They don't always agree with each other, and in some areas of the United States, neither is truly in tune with the local markets. We suggest, skeptical fellows that we are, that you go with the lowest figures. Both sites also include a full range of links to car buying services and insurance quotes.

This is only a sampling of the sites that we found interesting. An evening spent surfing automotive sites is sure to turn up many more. We hope this gives you an idea about the informational riches to be found on the Web, and ratifies our claim that this is a good place to start gathering information if you are contemplating buying a car.

Insurance Cost

Usually, we buy the car and then we insure it. Insurance, and insurance costs, may not play a very large part in our automotive choice. Then we get that bill every six months (and doesn't it always seem to come as a surprise!) and scramble to find the money to pay it. We suggest you check the insurance cost for your prospective car before you buy. Costs differ significantly in some cases, with the actual cost differences dependent on several factors—including car make and model—that insurers use to derive their often-differing premiums. (We will discuss these factors in more detail further on.) Let's review the types of automotive insurance coverage, and what they cover.

Liability

The amount of liability that a car owner must carry is set by the state in which the driver lives. Only three states do not require some form of liability insurance: New Hampshire, Tennessee, and Wisconsin. The highest liability requirements are in Alaska and Maine, both set at 50/100/25. The lowest is Mississippi with 10/20/5. What do these numbers, given in thousands of dollars, mean?

Let us take Mississippi as an example. The first number, which equals $10,000, means that that sum is the maximum covered for one person, if only one person is injured in the accident. The number 20 indicates that the maximum sum paid for the total number of people injured in an auto accident is $20,000. The number 5, $5,000, is the total amount that will be paid for property damage occurring in the accident. Doesn't sound like much, does it? It isn't. When you consider the costs of medical treatment today, and the cost of your average new car (about $20,000), then it appears that states like Maine and Alaska really do have more realistic liability requirements, high though they may seem. Because if your liability goes over the amount you are insured for, you are responsible for the rest. Yes sir.

This raises the question of how much liability you should carry. There is nothing that says you have to buy only the minimum required by your state. In fact, that decision could prove to be unwise. This is, of course, a judgment call. We suggest you review your circumstances in life and ask yourself, "How much do I have to lose?" not just, "How little can I get away with paying?"

Roy and Davidson both live in Florida. Florida is unusual in that the only legal requirement is property liability, not injury liability. Florida is, however, a "no fault" state, and, like all "no fault" states it insists on Personal Injury Protection (PIP) for all car owners, though only to the amount of $10,000. States vary widely in their PIP requirements. In Florida PIP pays for 80 percent of medical expenses, and 60 percent of lost wages, but these percentages can also vary from state to state. Almost any injury having to do with a car is covered, even getting your thumb slammed in the car door!

The standard (though not mandatory) policy in Florida is 10/20/10. Davidson, a cautious soul, carries 250/500/100. Roy is a bit more daring with 50/100/50. Neither of them enjoys spending money needlessly,

HAVING YOUR CAR STOLEN!

Some cars are more popular with thieves than others, but all cars are candidates for theft, from the fanciest foreign imports to the lowliest fish truck. It depends on what your car thief wants it for. It's not always to sell. Sometimes, thieves steal your car to use in a robbery, to transport drugs or guns, or to serve in some other crime; sometimes they steal your car for convenient transportation, or, perhaps, just for fun.

Car thieves can afford to be bold, because the process of tracing stolen cars and catching car thieves is cumbersome. And cars are pretty easy to steal. Anyone who has seen the speed at which a AAA road service technician, with the appropriate simple tool, can open up a locked car, will never feel secure about his or her car again—with good reason.

What's the best way to prevent your car from being stolen? Roy likes a steering wheel cuff called "The Club." He says the best way to install such a product is to install it with the keyway facing the dash, not facing the driver. Since the keyway is the most vulnerable part of the Club, making it a bit more inaccessible to a thief will give you that added bit of protection. Steering wheel cuffs will prevent a thief from driving off with your car, but they are not foolproof protection. The car can still be scooped up and spirited away with a tow truck or car carrier in an alarmingly short time. Davidson recalls a sight on Manhattan's Lower East Side: a car completely stripped right down to the chassis, sitting on bricks—no engine, no seats, no body parts, no radiator. The steering wheel was still there and—oh yes—so was the Club, firmly in place!

Other products have come on the market in recent years. One of the more interesting ones is LoJack, which places a tracking device in your car. If and when it is stolen, the police can activate it and locate your car, assuming that they have the equipment to do so. It isn't available in all communities and it isn't cheap—the basic package is about $600. It depends on quick action by the police, which is not always forthcoming. But it's clearly a step in the right direction.

Parking your car in a supervised lot makes sense, but it's not always practical, and it can be expensive. Common sense will tell you not to leave your car for long in questionable neighborhoods—not more than thirty seconds in some neighborhoods! But, by and large, you have to take your chances on the street.

What happens when your car is stolen? Davidson actually had the experience of having his car stolen off the street, from downtown Broadway, in New York City. It was a nightmare. Fortunately he had unloaded it before locking it up, so his loss was mainly the car itself and a few tools. First, even though he was legally parked, he had to check out whether it had been towed by some city agency. Since his theft took place on a weekend, this was not easy to do. He had to get a list of phone numbers for the various city car pounds from the NYPD, call each one (two of which never did get around to answering the phone) and, by a process of elimination, learn that it had not been towed. But when he tried to file a complaint at the police station he ran into further snags. They wouldn't let him make a complaint unless he could prove he owned the car. That meant a title, or the original registration. Well, the registration was in the car, of course! And no, a faxed copy of the registration wouldn't do.

And the title? Waaay back in Florida, with the bank that finances the car. Holy cow! Not only did the lucky thief get a great car, but also he or she could drive it with impunity on the street for two whole weeks before the owner could file a complaint!

So maybe it's a good idea to carry your car registration in your wallet, particularly if you are visiting one of the car theft capitals.

The National Insurance Crime Bureau (**http://www.nicb.org/services/top_stolen_cars2.html**) has some interesting statistics about the most frequently stolen cars by geographic area. Toyotas and Hondas are way up there, certainly. That makes sense; the most popular cars with thieves will be the most popular cars with regular people. But it's interesting to note that different cars and trucks are popular in different metropolitan areas. In Dallas and Tucson, Chevy pickups are the preferred stolen car, while Albuquerque favors Ford F150 pickups.

but both know that anything can happen in today's litigious society. Maybe car insurance is not the area in which to skimp. Hey, in Canada 100/300/50 is mandatory nationwide!

Not all insurance companies vary the rate for liability according to make and model. State Farm, for example, does not.

Collision

Collision insurance pays for damage repair or, if the car is considered totally wrecked by the insurance company, cash value—and sometimes replacement value—of your car if you have an accident and it is your fault. It is expensive, forming a major portion of your insurance costs if you choose to carry it.

Collision is smart when you have a new or almost new car. It is absolutely required if you are financing or leasing. But if you've paid it off, and as the cash value or replacement value of your car depreciates, you might want to consider whether it is worth the extra money to you. Again, this is a judgment call. Roy's program is to finance a car for five years, and take the collision off when the car is all paid for. He no longer pays collision on his '87 Taurus, for example, which has a replacement value of about $2,000.

Davidson still carries collision on his paid off '93 Integra, which has a cash value of about $6,000 and a replacement value of about $8,000. Some people might say this is pushing it, but if Davidson were to suffer a total loss of the vehicle, he would be hard put to replace it.

Comprehensive

Comprehensive covers your car from damage or loss from floods, storms, fire, and theft. If a stone thrown up by a truck on the highway breaks your windshield, it will also cover that. Comprehensive covers all damage or loss not connected with an automotive collision. Typically, comprehensive insurance is about a third the cost of collision, except in places like New York and New Jersey, where it is not unusual for them to be about the same.

How the Insurers Look at Your Car

Certain factors that the insurers use to calculate your insurance costs are not in your control. Your driving record is in your control, certainly, but at the point of sale it is a fixed element. You are who you are. If you have a history of accidents, moving violations, DUIs or other derogatory information on your record, you will have to pay the price. Your sex, age, and marital status are also factors not in your control.

Other factors, while theoretically in your control, are still not usually a practical consideration when buying car insurance. Location is one such factor. The lowest insurance areas in the country are North Dakota, Wisconsin, Iowa, Idaho, and Maine. These highly rural states seem to be full of careful, considerate drivers, who rarely damage their own cars or anyone else's. Too bad we can't say the same about New Jersey. Telling the insurer that you live in New Jersey will certainly set the calculator running. The Garden State has the highest premium rates in the nation, with an average annual premium for liability, collision, and comprehensive of about $1,300. If low-cost car insurance means a lot to you and you live in New Jersey, maybe you should consider a move. Don't move to New York, Washington, D.C., Connecticut, or Rhode Island, though. They are almost as bad as New Jersey.

That said, one factor they certainly do consider, and one over which you have absolute control, is the make and model of your car. To give you an idea of just what is involved when the insurance company looks at your car, we have chosen some representative cars from the years 1997 to 1999. We don't claim that they represent every type of car, but these are popular vehicles, and they may give you some insight about your car, or the car you are interested in buying or leasing. The information that we address in analyzing these cars ought to be of interest to you as well as your insurer. When we talk about a car's bodily injury or theft rate, and cost to repair indices, well, these are very important issues. After all, you and your family will be in the passenger compartment.

In order to gather this information we consulted insurance company executives from State Farm, Geico, and Farmer's Insurance Exchange. Another very important source for us, and for you, is the Web site of Highway Loss Data Institute, mentioned earlier, which is consulted by many insurers. This non-profit organization, financed with contributions from major insurance providers, maintains very complete statistics on bodily injury risk, repair costs, and theft loss. It is these statistics we are quoting in our discussion of these popular cars. Here is what the statistics mean: the bodily injury statistic represents the frequency of claims filed under Personal Injury Protection (PIP) policies; the collision and theft indices are derived from average loss payments per vehicle year.

Watch out for the theft index, though. A lower-than-average loss percentage doesn't necessarily mean that your prospective car is not attractive to thieves. What it means is that the average claim paid on that model, a Honda Accord with an index of 32 percent below average, say, is way less than that on a Mercedes S class, with a theft loss index of 901 percent above average. Both cars are popular with thieves, but you could buy five, maybe six, Honda Accords for the price of one Mercedes S600. And you might recover the Accord, but it's not likely you will ever see your Mercedes again. Now, then, let's look at our four selected vehicles.

BMW 525i, four door

This is certainly a luxury car by the standards of guys like us. And it's not cheap at 35 grand a pop. It's not cheap to insure, either. Your chance of a personal injury claim, however, is 30 percent below average with this solid car. But wait a minute! Should you be so unfortunate as to get into a crash with this vehicle, you can expect the repair costs to be about 41 percent higher than average. The theft index on this baby is 139 percent above average. That seems high, though not as high as a nearly new Acura Integra at 886 percent above average or a Mercedes S class, at 901 percent. In all fairness, though, those numbers are high not because they are so often stolen, but because they are expensive cars and when they are stolen they are almost never recovered. They drive them right onto the boat!

Chevrolet Cavalier, two door

This popular coupe is fairly inexpensive at about $13,500 and, with almost thirty miles per gallon (mpg) on the highway, it's not bad on fuel economy, either. This is not a good car to have an accident in, though, with respect to injury. The personal injury index is 29 percent above average. The collision index is 12 percent above average, too. Theft claims are 47 percent below average, however, which mainly reflects the low price of the car, not its unpopularity with thieves.

Chrysler Concorde, four door

This large solid family sedan was priced between $22,000 and $27,000 when new, depending on model. It's injury index is good, at 31 percent below average. Collision loss is 20 percent below average and the theft loss rate is 44 percent below average. Fuel economy isn't bad, either—about twenty-seven mpg on the highway and nineteen mpg in the city. This car won't cause any heads to turn when it rolls down the street, but it has good numbers.

Honda Accord, four door

A very popular mid-sized family sedan, this car is known for reliability and good handling. It comes is several trim lines ranging in price from about $17,000 to $25,000. Just a pop, 5 percent, above average on the injury index, 14 percent below average on collision loss. Theft loss rate is 32 percent below average. Again, this does not mean it is not a popular car with thieves. It is very popular. But it is not a very high-priced car and when stolen, can often be recovered, as opposed to a Mercedes S Class or BMW or other expensive luxury sedan, which are almost never recovered, particularly if stolen near a seaport.

Ford Explorer (SUV)

This mid-sized sport utility vehicle in the $30,000 range remains popular even though it has had a touch of negative publicity in recent years. Personal injury index is low, at 32 percent below average. But then, most of the big, heavy SUVs are fairly safe for the folks in the passenger compartment. We will not ask about the folks in the other car. Let's hope they were riding in SUVs as well. Repair claims are below average by 17 percent, too, though again, one wonders how the other fellow did. The theft index is just 2 percent below average.

Toyota Camry

One of the most popular family sedans in America, the Camry comes in a range of trim lines and ranges from $17,000 to about $26,000. The personal injury index is 3 percent below average, while repair claims are better still at 14 percent below average. The theft index is only 2 percent below average however. This car is very popular with thieves as well as the general population.

This '94 Pontiac Transport minivan still looks new at 163,000 miles. It recently came out of service as a company fleet vehicle and is now serving a new owner.

Insurance Companies

Buying insurance on the Web or over the phone from a discount insurance company like Geico or Progressive might save you money if you have a good driving record. It's not a bad option—their insurance is just as good as anybody's. There's a tradeoff, though. Traditional insurance companies like State Farm, Allstate, or Nationwide might cost a little more, but they have networks of brokers all over the country and are able to give you service that is a bit more personal, and maybe even a little more rapid. If you get into trouble away from home, a call to a local agency that deals with your insurance carrier could save you some heartache. At the very least it will give you someone local to talk to about your troubles, and someone who is likely familiar with area repair shops. It's a point to consider.

Other Cost Factors

Cars are a compromise. There is no ideal car; your job is to find what is close to the ideal car for you. High ratings for safety, fuel economy, comfort, and passenger capacity, low repair and maintenance costs, greatest ease of handling and parking, do not all add up to the same car, unfortunately. So you have to decide what qualities are most desirable for you. This means asking yourself some important questions. We can help by leading you through some of the more obvious issues.

Fuel Economy

If fuel economy alone is all you're looking for, then take heart. According to a report of the Commission on Engineering and Technical Systems, back in 1988 they made just the car for you. How does 6,409 miles to the gallon sound? And while you are celebrating that idea, how does fifteen miles per hour sound? And you'd better lose some weight, because the passenger load was less than one hundred pounds. I don't think the test course included any hills. Pretty silly, eh?

For your information, though, the model year 2001 gasoline-powered car with the absolute best fuel economy is the Honda Insight with sixty-one mpg in the city and sixty-eight mpg on the highway. Not bad. This interesting little two-seater is an innovative combination of one-liter gasoline engine with a self-charging electric motor assist. It does have a huge battery pack, and they are expensive to replace.

There are other choices that are also not bad. The new VW Beetle with the diesel engine will give you forty-nine mpg on the highway, and the Honda Accord will give you thirty-two mpg—a pretty good compromise considering what else you get with that ever-popular car.

The United States Department of Energy lists the best and the worst cars in terms of fuel efficiency only at: *http://www.fueleconomy.gov/feg/ bestworst.shtml*. There are some surprises. Check it out!

Safety in a Crash

Several organizations conduct crash tests, and before you buy any car you should evaluate its crash-test results. Those results may not be the sole determining factor in your decision, but you should be aware of them and of the particular risks involved with the car you finally choose.

Obviously, if you are transporting a car full of children on a regular basis you are going to pay quite a lot of attention to crash-test results. You don't want a one-star vehicle. We suggest the same if you are the kind of guy who likes to drive at speed on backcountry roads, or if you spend a lot of time driving the interstates. You just never know. You might be the best driver in four counties, but some nincompoop could still pull out at the wrong time, or T-bone you at a red light. You are at other people's mercy when you are on the road. We can't think of a single reason not to pay attention to crash tests.

The National Highway Traffic Safety Administration does crash tests and has a quite complete site at: *http://www.nhtsa.dot.gov/cars/testing/NCAP/SaferCar2000*. Another good site is that of the Insurance Institute for Highway Safety. Their crash-test results are at: *http://www.carsafety.org/vehicle_ratings/ratings.htm*.

Parking

If you live in the suburbs (as more than half the people in the United States do) and park in a company parking lot at work, then parking, except for trips downtown, is not such a problem. If you live in the city, though, you are a bit more pressured. A parking garage can be very expensive. In New York, you could easily pay $300 per month to park your car. Size of car makes a difference in many urban garages.

Another factor is the ease of parking itself. Both Davidson and Roy have been caught short by cars with bad turning radii. A short turning radius is the key to easy parking and to making quick, effective U-turns. It can save you much embarrassment and improve your safety when tight turns cannot be avoided.

So can good visibility. You want to be able to see in order to correct your parking maneuvers. Ideally, you should be able to see the corners of your car, but they don't make many cars that way. Nonetheless, you should have a good field of vision all around the car, with few or no blind spots.

A STATUS-FREE CAR?

Wouldn't it be great if a truly status-free car would appear on the market again? We can only recall two in our whole automotive history. We think the Model T Ford qualifies. Variously priced from $265 on up, it was a true "peoples' car" owned by millions of Americans in all walks of life, rich, poor, and middle class. Ford made 15,000,000 of them between 1908 and 1927—a long run for one model.

We did not see its likes again until the mid-1950s, when Volkswagens began to appear on American highways. Back then, we called them "Bugs," not Beetles. They caught on slowly, at first. In 1957 Americans bought 54,000 VWs, but in 1967 they bought 340,000. If we consider that the model began in 1945 (actually there were prewar versions of the VW) and that the Bug is still being manufactured in Brazil and Mexico—well, that is a long model run. In fact, the twenty millionth Bug rolled off the line in Puebla, Mexico, in 1981. It was the most popular car in the world, and its record still stands.

The great thing about a Bug was that ownership in no way defined the status of the driver. He or she could be a multimillionaire or a poor student, or anything in between.

Why was it so popular? Well, the American cars of the 1950s were not called "Detroit Dinosaurs" for nothing. They were huge: eighteen, even twenty, feet long, with big bench seats, huge hoods and trunks, and covered with glittering chrome. Gas mileage? Twelve mpg in the city, seventeen mpg on the highway wasn't bad for those days.

The VW was the absolute antithesis of the standard American car of the '50s. To begin with, the engine was in the rear! It was small, economical to run and maintain, cute, fun to drive, and cheap. And it would have been a pretty safe car if all the other cars on the road had been VWs, too. But Americans lived dangerously in those days—smoking, drinking, and eating saturated fats. Those were the days!

Warranties and Other Options

Warranties and Extended Warranties

All new cars come with manufacturer's warranties, typically for three years or 36,000 miles, whichever comes first. The important thing to remember about warrantors—car manufacturers and extended warranty companies alike—is that, like insurance companies, they are in the business of *not* paying. You almost have to be a lawyer to understand the fine print on warranties, but let us simplify it for you. If a part, or part group, is not specifically covered in the warranty, and if you have not followed, and documented faithfully, the maintenance schedule suggested by the manufacturer, or if you are one day out of the warranty period, then they will do their level best to avoid paying. That goes for all car warranties. Also, car manufacturers will guarantee only those parts that they installed in the car when it was manufactured. They do not cover, for example, dealer installed options.

Still, warranties are not totally useless. Parts do break down, or are sometimes carelessly installed. When they do or are, and all the other conditions are right, you may be able to implement your warranty without too much fuss. Also, sometimes a manufacturer will repair something that is out of warranty on mileage, but is still in the warranty period. They will not volunteer to do this but they may do it, or at least share the cost with you, if you demand it loud enough.

Extended warranties do basically the same thing that manufacturer's warranties do, but are often even more specific as to just what they cover. They can and do offer extra bells and whistles like roadside assistance, and car rental allowances so you can rent a car while yours is in the shop.

Extended warranties are high-profit items for dealerships. You would be lucky to get out of the finance office without being made to feel you have to have one. Resist. Buy no warranties at the dealership! For one thing, do you want to pay 8.5 or 9 percent interest on an extended warranty by tacking it onto your auto loan? Certainly not! Davidson feels that the best place to buy an extended warranty, if you really must have one, is directly from a warranty company. There are many such sites on the Web. Just go to your search engine and type in "automobile warranties." There are lots of varieties of warranties out there. For exam-

This 1992 Jeep Wrangler is on its second owner, with just over 145,000 miles. The current owner has account-ed for nearly 100,000 of those mostly trouble-free miles. This Jeep has proven worthy on both 450-mile trips and during a daily 32-mile commute. Through meticulous care, the owner hopes to add at least another 100,000 miles before "retiring" the Jeep to weekends only.

ple, some are good nationwide, while others are only good at one shop or service department! We'll talk some more about this in chapter two.

Be careful, however. Warranties are only as good as the companies that back them.

Weighing Your Options

All new cars come with available options, although, more and more these days, the list is small. Car manufacturers produce their models in various trim lines, and each line comes with a list of features that includes more than the line just below it. Actual options can be as few as foglights, say, and a choice between manual and automatic transmission. Practically speaking, unless we order our cars from the factory, new cars, like used cars, come with the options already installed. Other options may be available, but these are often installed at the dealership and are not covered by the factory warranty.

That said, when you are looking for a car that you are going to spend a good part of your time in, we recommend that you consider a car that has some "smart options." What we call "smart options" are those that really do augment the car's economic and comfort value, and are well worth considering. Three hundred thousand miles is a long way to go in a car, and your personal comfort is a very important consideration when choosing a car and options. We like options such as power windows, power seats, interval wipers, cruise control, and keyless entry. Leather or cloth interiors tend to sell better and be more comfortable than vinyl, though vinyl lasts and lasts. Leather requires some additional maintenance, but it, too, will last a long time if you take care of it, and it does enhance comfort and resale value. Both Roy and Davidson fully appreciate a comfortable car.

What Must Your Car Do for You?

Everybody uses their car to go to the store; most use it to get to work. But many people also have special uses and requirements for their cars that must be considered when they choose what car to buy. Some people tow boats—unless you are towing a kayak, a little four-cylinder econobox will not do. If you tow a boat or anything heavy you will want a substantial car with a powerful, well-maintained engine, and brakes in tip-top condition. Towing puts a real strain on car and engine.

Some need to carry their tools around in the car, while others need to move boxes and such from time to time but don't want a truck or large van. A minivan might be a good choice for those folks. Minivans have some great features. They will haul people, small animals, a big toolbox, and some appliances, all the while providing a long-distance ride that is certainly the equivalent of most cars in terms of comfort. You can even sleep in one if you have to.

People with young families have special concerns about the safety of the cars they drive. A car that rates one star on crash tests is not going to be their choice. Single people may look at their car as a component of their social lives. A convertible or sports coupe might be just the ticket. A person whose job keeps them constantly on the road—traveling salespeople, journalists, country veterinarians—will place a very high premium on fuel economy, personal comfort, and reliability. And of course, there are those who want their car to exude a certain image or personality.

Physical characteristics should also be a prime consideration. Roy is six feet, four inches tall, and is not going to fit too well into a compact car unless it has a sunroof. Davidson suffered an acetabular fracture some time ago, and is therefore very choosy about driver seat comfort, particularly in the lumbar area. Some people just don't fit with certain cars. If you have long legs and you car pool every day, then you sure don't want a car with little or no legroom in the back seat. You want to be able to put that driver's seat back without crushing the legs of your fellow employee. Roy likes Corvettes, but he can't drive one because he's so tall that his eyes are at the level of the top frame of the windshield. He says, "I literally have to fold up in the seat to drive a hardtop 'Vette, and it's strictly top-down with a convertible!"

Credibility: Is Your Car Part of Your Image?

It's no secret that we are often judged by our cars. Depending on who is doing the judging, it can be amazingly sophisticated, starting with the type of car, and working right down to the make, model, and year. Car dealers, bankers, prospective employers, country club layabouts, con persons, and inner-city pimps are all experts at judging just where you fit into the scheme of things by looking at your car.

You can't avoid it. But you can play the game your own way, as many people do. Quite a few rather well-off descendants of old New England families tend to prefer older inconspicuous cars—not that they are totally anonymous. A low license plate number is a dead giveaway in Connecticut, where early numbers can be passed down through families. And some of the richest guys in Texas drive around in old pickup trucks, we are told. And, of course, those of a certain age will never forget Jack Benny's old Maxwell.

In our opinion, having a clean, well-maintained, and well-loved car of any age or model says something very positive about its owner. It suggests that he or she is frugal, not a slave to fashion or status, or to the idea that you have to spend a lot of money to reinforce your self-worth.

Arriving at Your Dream Car (The Ideal Car for You)

Car Evaluation Reports

When you get to a certain point of car-buying readiness you will be reading car reviews, test-drive reports, and such information as daily fare. For new cars, this is really all you have. After all, when a car is new, the jury is still out on it. A few years down the road we will be a bit more aware of its strengths and failings.

That is another advantage used car buyers have. Their prospective car already has a history—at least, the model does. Where do you go for such information? A classic reference for used cars is Edmunds. For years, Edmunds has evaluated used cars in their annual guide; now they do it on the Web, and it's free. This good site will lead you, step by step, model, year, and trim line, to an evaluation of the car you are thinking of in terms of safety, reliability, performance, comfort, and value. Ratings are given on a scale of one to ten. These evaluations are done with a pretty broad brush, but a lot goes into them and they do give a good sense of one car's advantages or limitations over another. Sometimes the information includes a review, what features were new for that particular car that year, and links leading to other information about the vehicle. It also includes their assessment of the trade-in, wholesale, and retail price. Pretty good. We would certainly not buy a used car without looking at its Edmunds evaluation. (See Edmunds at: *http://www.edmunds.com.*)

Or without checking it at Intellichoice. This interesting car evaluation site estimates just what owning the car you are thinking of buying will cost you over the next five years, rated in terms of depreciation, finance costs, insurance costs, fees, fuel, maintenance, and repairs. Makes and models only go back to the year 1997, but still, it's a very handy reference if you are buying a recent model used car. Roy points out that "many '97s are carry-over models, anyway. In most respects, they are almost identical to previous models of the same body style. For example, the Ford Taurus was restyled in 1995, and the basic car hasn't changed much since. It has just been enhanced each year." Intellichoice also comes up with a list of best values in both new and used cars, by category and price range. (See Intellichoice at: *http://www.intellichoice.com/ic2.*)

There are other car evaluation sites worth visiting also. Good old *Consumer Reports* does a very good job with used cars, rating models with the familiar red to black dots for a pretty lengthy list of factors including: engine, cooling, fuel, ignition, transmission, electrical, air conditioning, suspension, brakes, exhaust, body rust, paint and trim, integrity, hardware, and overall reliability. Their reports are not too detailed, but they are not afraid of making sharp criticisms, and their evaluations seem right on the money. Of particular interest are their categories "good bets" and "reliability risks" featuring the best and worst of used cars according to reliability history. Yep, Davidson and Roy would certainly check those out before we plunked our money down.

Unlike the other sites we mentioned, the *Consumer Reports* Web site is not free—you have to subscribe, and it costs about $24 per year. It is well worth it, though. Not only is it a big help in evaluating cars, but also in evaluating tires, batteries, and other auto accessories. They also have various electronic subscriptions available on their Web site. (See *Consumer Reports* at *http://www.consumerreports.org.*)

In Conclusion

No, we can't tell you what car to buy—that decision has to be yours alone. As we have said before, today's cars are better, in general, than the cars of a decade ago. Customer demands, government regulation, and—probably even more important—foreign competition, have made them so. So you have a pretty wide range of cars to choose from.

What we have done in this opening chapter is give you a list of things to think about when you choose a car, and a list of places to go to get the information you need. Don't be limited by our choices. These are good places to start, but information (and misinformation) can come from the most unlikely sources. Always be open to new information, but always be skeptical until you have checked it out yourself.

We tend to evaluate cars in terms of safety, reliability, economy, suitability, and comfort. We give each factor about equal weight. All are very important—skip one and there will come a time when you will be sorry you did.

TAKING POSSESSION

We could write an entire book about the finer points of buying or leasing a car, or the "showroom experience," if you will. In this book, though, we want to guide you through the whole car ownership experience, not just closing the deal. So the information in this chapter is necessarily limited to what we feel you *have* to know in order to buy or lease a car. That said, let's begin our discussion of how to actually take possession of the perfect (for you) car or truck.

To Lease or to Buy

There are only two choices available if you are to drive away in the car or truck you have just picked out. You must pay for it through a leasing agreement or buy it outright. In order to know which to choose you must know what terms are available and then decide which option gives you the most advantage.

We know that you understand the difference between buying and leasing, but bear with us for a moment. For us, the difference between buying and leasing is this: When you buy you are paying for the entire expected life of the vehicle with traditional financing; when you lease you are paying only for the portion of a car's life expectancy that you actually use.

The very premise of this book is that today's car or truck can be successfully and happily driven for 300,000 miles or more. For most of us, that translates to about a fifteen-year life span at 20,000 miles per year. At the rate of 15,000 miles per year, that's a twenty-year life span! That, we hope, really brings home to you the idea of getting the maximum economy out of your car!

The plain fact is that a reliable fifteen-year-old car will, if you pay even the most cursory attention to the speed limits, get you where you are going just about as fast as a new Porsche. In light of that irrefutable fact, the decision to replace a vehicle is based mainly on a series of very personal value judgments. And every person's decision is unique.

For any number of reasons, many new car purchasers keep their vehicles for only three to five years and then get another new one. And some drivers buy maybe five cars in a lifetime and run them until the wheels fall off. Obviously, different folks have different ideas about the "useful life span" of a vehicle. That being true, the payment plan that makes the most sense for one owner will not be best for another.

Financing a car is all about depreciation and resale value. Note that the owner of a new car takes a substantial depreciation "hit" in the first few months, regardless of the payment plan. At full resale value, the average car is worth 85 percent of its original suggested retail price at the end of six months, and about 65 percent as a trade-in. After the first six months, this depreciation rate slows to 10 percent per year, with full resale value—as published in the used car guides, for example—taking a slightly sharper dip than trade-in value. You can expect a car's trade-in value to average about 75 percent of its full resale value after the first year has passed.

The Finance Equation

To begin our comparison of leasing and buying, let's look at an example of a new car purchase with traditional financing. To keep the calculations simple, we will assume that you are going to buy a new vehicle with a total financed price of $20,000, including tags and taxes, but not including insurance, extended warranty, or maintenance costs. We'll

Typical New Car Financing — Depreciation versus Equity

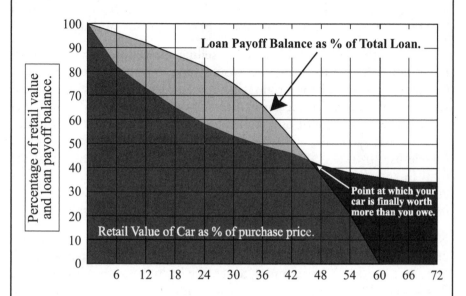

Here's another way of looking at it.

Months of ownership

	0	6	12	18	24	30	36	42	48	54	60	66	72
Loan Payoff Balance as % of Original Loan	100	82	73	65	58	53	49	46	41	38	36	34	34
Retail Value of Car as % of Purchase Price	100	96	92	87	82	75	66	52	37	21	0		

But, whichever way you look at it, here's a good argument for not buying a new car, at least if you use conventional financing. You're upside down with it right at the git go, and will stay that way for almost four years!

Figure 2.1—Until the lines cross, at about 46 months in this example, the car buyer is "upside down," owing more on the auto loan than the vehicle is worth. Note the tremendous drop in value in the first six months!

also assume that this new vehicle will be worth $10,000 at resale in four years. If the vehicle were financed at 11 percent for forty-eight months, the monthly payments would be $517. So, at the end of forty-eight months, you would have paid $24,816 for a vehicle that is worth $10,000—*if* sold for full retail, not as part of a trade-in deal!

This means that if you did succeed in selling the vehicle privately for its full retail value—$10,000—you would have effectively paid $14,816 (about 75 percent of the original borrowed amount) to use it for four years. But if you traded the car in after four years, the dealer would likely pay an actual value of only about 65 percent of retail, $6,500 in this case. In that scenario, you would have paid $18,316 for four years of vehicle use. In other words, you have paid out over 91 percent of the total financed price to use the vehicle for only about 25 percent of its useful life expectancy. The real benefit of owning this car will likely go to the second owner, who doesn't stand to lose all of that value.

Now, suppose that the car you purchased has a low resale value, or you drive fifty thousand miles a year, and now it is worth less than you still owe on it. And you are planning to replace it! In the auto sales business, this is known as being "upside down" with your car, and it may be a recipe for financial woe, especially if repeated. In this sad scenario, the purchase price you will be financing for your new vehicle will have to include the amount needed to pay off the old one. You could end up with a loan for the full purchase price, even after making a substantial down payment—or even worse, trying to finance more than the car is worth, even when new! Even if you are not planning to buy or trade, there is a certain bitter feeling that goes along with owing more on your car than the car is worth.

If you normally trade your car in for a new one every five years or less and plan to continue to do so, leasing may be a way to avoid the type of loss that we just described. In fact, consider this: you may have been leasing a car all along and not have even realized it. If you replaced a car before it was paid off, you probably never even saw the title to it. In effect, you were leasing the car from the bank, which actually owned the vehicle all that time! In that case, you may have been better off to go with a "real" lease agreement, rather than traditional "financing to own."

Not necessarily, though! Leasing has its advantages, but there are also some serious pitfalls. But first, let's talk about the advantages.

The Advantages of Leasing

It is with good reason that most all new car dealers' commercials cite two numbers for each car they advertise. "Drive this car today for only $18,995, or only $325 per month!" It sounds like they are describing a single deal, but that second number, the "per-month" figure, is almost always a lease payment and is lower than the monthly payment would be to purchase the car. That is one of the advantages of leasing—a lower monthly payment for the same car.

To extend that thought a bit, it may be possible to drive a more expensive car at the same monthly cost. By leasing, you are paying only for the depreciation the car sustains during the term of lease, instead of paying the entire purchase price. In addition, little or no down payment is generally required with a lease. At the end of the lease, there are usually several options that may be exercised. You can generally opt to:

- Buy the car for a discounted price, resell it, and make a profit.

- Buy the car and keep it. After all, it is now a used car that you already know in detail, from the service history to the fuel economy. No surprises here!

- Walk away from it and get a new car with no cash outlay. This would seem to be the option that most leasing customers have in mind. Assuming that the lower monthly payment is a choice rather than a matter of necessity, you would also have the opportunity to invest the amount you would otherwise be paying out every month in an income-generating account.

This concept sounds good if you plan to trade cars on a regular basis, but all leases are not created equal. There are variables that you must consider in order to determine if a particular lease program is for you. So take the time to investigate the types of leasing available and decide what meets your needs. Remember, *you do not own the vehicle at the end of a lease!* If, at the end of the four years we used in our example, you want to keep the car, you must buy it in the traditional way.

Two Types of Leases—A Bad One and a Good One

There are two basic types of car lease: *open-end* and *closed-end*. The keyword here is "end." Both terms have to do with what happens at the end of the lease. With an *open-end lease,* you are liable to the leasing company for the difference between the projected resale value of the car as stated in the lease at the time of signing (called "residual value") and the actual market price of the car at the time the lease expires.

This type of lease typically has the lowest monthly payment, but generally lists the residual value (the projected value of the car at end of lease) somewhat high. If the actual value of the car turns out to be lower than the stated residual, the lessee (you) may have to come up with a substantial amount of cash at the end of the lease. Obviously, there are lots of factors that can influence the resale value of a vehicle three to five years down the road. A recession could really make inroads on used car prices. You might acquire some scratches or dents, or your dog might have done something in the backseat. Or you might, if you travel a lot, have put more miles on the car than the leasing company estimated. You better have it detailed before you bring it in! If the market value of the car you are returning to the leasing company is less than the projected value in the lease, you will owe some money. Got a crystal ball? You'll need it!

The *closed-end lease* is essentially a walk-away lease, meaning that the lessee can opt to purchase the car or simply walk away without having to pay, and it comes in two varieties, *fair market* and *fixed residual.* With a closed-end, fair-market lease, the purchase price is based on the actual fair-market retail value of the vehicle at the end of the lease term, or the residual stated in the lease agreement, whichever is *greater.* The leasing company may elect to have the car appraised, but they will usually agree to sell it for the residual stated in the lease. It's their call.

A closed-end, fixed residual lease is also known as a "stated purchase option lease." With this type of agreement, a specific purchase price for the vehicle at the end of the lease term is stated. This is the only type of lease that guarantees the purchase price, and this protection may provide an opportunity to purchase the vehicle and sell it for a higher price, or it may provide a hedge against inflation if the lessee intends to keep the vehicle. The residual value will generally be set slightly higher than with a fair market lease, resulting in slightly higher monthly payments for this protection. Generally, the stated residual value should be about 30 percent lower than the actual retail at lease-end, unless economic conditions have caused unusual fluctuations in retail value.

Other Pitfalls

In addition to considering the differences in the way the lease agreements may be written, there are some pitfalls to watch for in any lease. These are:

- Stipulations regarding the condition of the car at lease-end: Virtually all leases state that the car will be returned in no worse than average condition, with no more than normal wear and tear. Anything that needs to be replaced or repaired can result in additional charges. Items that commonly run up turn-in charges are worn tires, exhaust leaks, worn brakes, excessive door dings or chips in the paint, dirty or torn upholstery, and missing or damaged trim or accessories. As we mentioned before, getting it detailed before you take it in might prove cost effective.

- Excess mileage charges: Leased vehicles have a mileage allowance written into the agreement, usually fifteen thousand miles or less per year, depending on the lessor who is writing the lease. If the vehicle is driven more than the allowed number of miles, a charge is levied at lease-end, generally at least fifteen cents for each additional mile. That can really add up, but it is only an issue if you do not purchase the vehicle. If the lessee buys the vehicle, the leasing company doesn't care how many miles it has on it. Note that expected excess mileage can be written into the lease agreement in advance (reducing the residual value at lease-end) but once it is done, it cannot be changed. If you drive fewer miles than you expected, you still pay for the mileage stipulated in the lease.

- Early termination charges: These may be built into a lease agreement, and are usually quite expensive, unless the lessee can find someone to take over the remainder of the lease agreement for the same monthly cost. Most lessors will allow this, but there will still be processing costs to change the paperwork. In short, if you have to terminate the lease early for any reason, you pay dearly!

- There are some Web pages which offer more detailed information about auto leasing:

    http://www.leasinghelpline.com
    http://www.lectlaw.com/files/cos17.htm
    http://www.hs.ttu.edu/ffp1370/chapter_seven/page6.htm
    http://www.dca.ca.gov/legal/l-6.html

A great looking 84 Buick Regal! This little beauty is still driven daily, after accumulating over 175,000 well-maintained miles.

Shopping for Car Leases

Leases can be obtained from three types of sources:

1. Automobile dealerships

2. Independent leasing companies

3. Service/membership organizations, such as insurance companies, motor clubs, and banks

Each source will have advantages and disadvantages in your particular circumstances, so take the time to shop around *before* you are ready to sign the papers and drive away! A good way to begin exploring is to look for options on the Internet. Type in the keywords "auto leasing" using your favorite search engine and you will find dozens of leasing companies that will provide quotes and details. This can save you lots of time in your quest for information, especially at the beginning of the process. If you do not have your own personal computer with Internet access, you can likely arrange to use a PC at your local library to go online and begin exploring.

Traditional Financing

Now, let's go back and see what happens if we finance our car in the traditional way and do *not* sell or trade it at the end of four years, but, instead, keep it in the driveway. If you elect to keep a car for its entire life span of fifteen years or more, and pay it off in forty-eight months, you will have a vehicle that is "free and clear" for the remaining 80 percent of its life. When you make that last monthly payment, you own the car. This is where you begin to reap the benefits of buying rather than leasing. From here on, your monthly car payment is zero. Think about that the next time you are writing out checks for the monthly bills. That's one big one you won't have to write anymore.

Try to arrange financing before you buy the car. *We are really very insistent about this.* Financing the car in the dealer's showroom is very dangerous—for reasons we will elaborate on later. Before you even walk into the showroom, shop around the neighborhood banks, credit unions, and other lending institutions. And don't neglect the Internet. Simply type in keywords such as "auto financing" or "auto loans" and you will find numerous Web sites to explore. The beauty of Internet shopping is that it is quick and you don't have to commit to anything in order to get an initial quote. The Internet can give you an almost instant feel for the auto loan climate as it pertains to you!

While considering different financing options, be sure to look at different loan *terms* as well as interest rates. Sometimes, you can achieve a more advantageous interest rate, and shorter loan term, for nearly the same monthly payment. For example, reducing the term from sixty to forty-eight months will sometimes lower the interest rate to the point that it will cost only a few dollars more per month to buy the car. If you can stretch a bit and make the slightly higher payment, you will reap substantial savings from not paying that extra year of interest. You will also decrease your chances of getting upside down!

Once you are armed with a good sense of what terms and rates are available, you can negotiate financing for a particular make and model, with specific options, before you go to the showroom. This is the best way—way better by far than negotiating finance in the showroom! When you have an approved loan commitment, you can go to the dealership "already bought at the bank" (the auto sales department term for having your purchase money already arranged), so you can negotiate your best deal with the knowledge that you are getting a favorable loan. This is one way to gain an advantage when dealing with the salesperson—we'll talk about this some more a little later in the chapter.

By the way, this philosophy should be applied to used car deals as well, whether you are contemplating a purchase from a dealer or purchase from a car owner. If your credit is decent, you will likely be able to arrange a better deal on your own than most dealers will offer you on a used car. However, do keep in mind that most lenders will not give their most favorable financing on a used vehicle.

If you are looking at a very late model used car, take the plunge and compare the loan quotes for a new vehicle. You might be surprised at how close to each other the monthly payments really are! The term and rates may vary considerably based on the age and mileage of the car or truck being financed. If the value of the vehicle is relatively low, you may essentially end up with a "signature loan"—a loan that is not actually secured by the car. The bank is telling you that they don't really want the car back, even if you don't pay! The amount of financing that lenders are willing to approve for a particular vehicle can also help to provide guidance on whether or not the vehicle is fairly priced.

In any case, having your financing in hand before you buy the car is almost always better than doing your financing at the dealers. Of course we have all seen TV commercials from auto makers that advertise favorable financing options and incentives on certain models, usually "for a limited time." While these offers are genuine, they are *not without fine print*. To begin with, note the keywords "on certain models." Not only is that the case, but they may specify exact options, as well. And often those rates are based on a shorter loan term than usual, often thirty-six months or even less. These days, it is unusual to finance a car or truck for less than forty-eight months, and a sixty-month term is even more common.

However, if you can afford the terms and find the specified model acceptable, you may be able to take advantage of a great deal. Just beware: many customers will be lured into the showroom by these offers, find that there is no benefit for them, and be seduced into a much less advantageous financing arrangement and/or a more expensive car. The effect is a little like the old "bait and switch" technique.

So, Lease or Buy?

To sum up, you may be able to benefit from leasing a new car or truck if you intend to keep it for less than five years (even if what you would otherwise buy would be a late model used vehicle), or if you have a set monthly allowance for car expenditures. However, if you are content to drive an older car and plan to follow our advice to drive it happily for 300,000 miles, traditional financing is probably for you. It is also possible that a poor credit history, high debt-to-income ratio, or frequent employment changes may make leasing approval difficult to obtain.

As with any major purchase, the payment plan you select for your car or truck must be considered carefully and adequate time must be spent on evaluating alternatives. "Gotta have it now" is a mindset that has robbed many people of the opportunity to get the best deal or to pay in the way that is best for them. We will talk more about this as we help you decide where to buy or lease your car and how to get the best deal.

Figure 2.2—No, it's not an underwater picture of the Titanic—it's a rusted out floorboard on an otherwise good-looking '84 Ford Ranger. Roy discovered this after he bought the truck without having inspected it underneath (such a deal!). Lots of Michigan winters and an unrepaired water leak in the cab took their toll on this little truck. This damage was invisible with the floor mat installed. A new floor pan and body mounts had to be installed, more than doubling the "bargain" price!

Used or New?

You may recall our mentioning the fact that the largest chunk of depreciation on a new car occurs in the first six months of its life, and that the best value may go to the second owner if a car is replaced while still young. Like any other purchase judgment, *you* have to decide whether or not to consider a used vehicle. However, we are able to offer advice and guidance to help you develop your personal decision.

There are four ways to acquire a car these days, new or used. You can:

1. buy onsite from an auction

2. buy from an Internet auction

3. buy from an Internet Web site (most likely involves a dealer, sooner or later!)

4. buy onsite from a dealer, the "old fashioned way"

In addition to these methods, you can buy a used car directly from its current owner, through an ad or simply by responding to a "for sale" sign.

STOP

ABOUT ADVERTISED PRICES

Keep in mind that an advertised price is usually more negotiable on new vehicles than used, but it's a good starting point for either. The price quoted in an ad is likely for a specific vehicle at a specific dealer, as indicated by a stock number in small print at the bottom of the ad, and/or it may reflect the net price after certain incentives or terms are included. Do not fall into the trap of believing that the dealer who advertises they will beat "any advertised price" or "your best written deal" will necessarily give you the best deal. Remember, all they have to do to beat someone else's deal is to give you a few more dollars off (and then try to get it back as they close the deal with you!).

YOUR CANCELLATION RIGHTS

It is legally required that you be told of all your cancellation rights in your state and locality before you sign on the line. **Sign nothing binding until you fully understand these rights, whether buying electronically or in person!** It is also important to remember that, even after you have signed the buyer's order and given a deposit, the deal is not final or binding until the contract is signed by an authorized representative of the dealership, usually the sales manager. You can walk away whenever something does not feel right. You can even return up to three days later to walk away!

The obvious advantage of buying onsite is that you get to touch, drive, and inspect the actual car or truck you buy. Roy would never buy a new or used car any other way, because he always seems to choose a particular car based on rather subjective criteria (whether it "feels right") as much as on value, price, and ratings, and he likes to drive more than one car of the same make and model. He also admits to *enjoying* the whole negotiating process with car sales professionals and current owners!

Davidson feels the same way. Usually he knows exactly what he wants as to make and model when he decides to buy a car. Repair history, reliability, and safety-in-a-crash, for that particular make and model, are all big factors in his car-buying decisions. Yet like Roy, he would never buy a car without seeing it, driving it, and checking it out pretty thoroughly. He doesn't enjoy negotiation as much as Roy does, but being committed to used cars as he is, he knows that negotiation is part of the deal.

While we absolutely recommend that you not sign a sales agreement without having driven *that particular car,* we are quick to point out that there are advantages and disadvantages to each method of acquisition.

Buying at Auction

Buying a used car at auction, on the Internet or in person, can result in a very good deal—or a very bad one. Most "live" auctions, other than those held by local law enforcement agencies, are reserved for dealers only, but there are some public auctions to be found, and more on the Internet. We're not going to spend much time discussing this option, because we *do not* recommend it for most buyers! "What you see is what you get" best describes the auction experience.

Auction sales are usually final, and "as-is," with no warranty. This is why auction prices are often below wholesale. With only a few exceptions, you buy a vehicle that you know nothing about and likely will not get to drive before buying. Occasionally, a car or truck is sold at auction "with a drive," meaning that you get to take a short test drive and have the option to cancel the purchase, but this is the exception, not the rule. Suffice it to say that Roy, as well as some of his colleagues and relatives, all with years of automotive experience, have all been "burned" buying cars at auctions!

Buy It on the Internet?

There may come a time when you buy all your cars on the Internet, but we aren't there quite yet. Most folks still buy off the lot. You *can* buy a car on the Internet, though, and even if you don't close the deal there, there is no quicker way to check out the market. Most of the sites we mentioned in the previous chapter will certainly have links to Web-oriented auto dealers. We would say that the Web should be the first stop for the serious new or used car buyers.

The Web is also a great place to get a sense of which cars and trucks are likely to be deeply discounted and which ones are "hot models." The hot models are usually the newest and sportiest ones, which are so popular that they sell at or above suggested retail.

Once you have a sense of what the car or truck you want should cost, you can either remain on the Internet to make the deal, or use the information to begin negotiations with a dealer or owner. When using the Internet, you may be directed to a particular dealer (if the Web site is actually a locator), or you may be able to complete the entire deal electronically, up to the point of signing the sales or lease agreement.

One statistic we saw recently seemed to indicate that the average new car quote obtained on the Internet is about 6 1/2 percent higher than the "best deal you could get in the showroom." We disagree: that "best deal" figure is both subjective and elusive! There are so many variables that change the "best deal" dollar amount from one locality to another and one vehicle to another that we feel very uneasy about quoting such a statistic as accurate. In our experience, the lowest quotes obtained on the Internet for new and used cars almost always represent fair prices that you may or may not be able to beat at the dealership under given circumstances.

Buying electronically also avoids the hassle and time of door-to-door shopping. It's not a bad way to buy, especially if you are not a strong negotiator, or feel intimidated by the whole purchasing experience. If you do make the entire purchase on the Internet, or by phone through a buying service, your car may even be delivered right to your door, and your trade-in taken away, if that is part of the deal. You have the option to cancel the deal if you are dissatisfied for any reason, right up until you sign the paperwork and three days thereafter (a "cooling off" grace period required by law).

For Sale by Owner (FSBO)

If you are looking for a used car or truck, another way to avoid many of the aggravations associated with pur-

FACTORY REBATES

If a rebate from the manufacturer is available on the car or truck you wish to purchase, there are two ways to receive it. You can receive a check directly from the manufacturer, or you can have the rebate applied against the total financed price of the vehicle as additional down payment. The second option is usually best, because you will save interest expense by reducing the amount to be financed. That way, the savings will be substantially greater than the actual rebate amount. The rebate, or other sales incentives, may already be figured into an advertised price, or it may be offered as an alternative to another "discount," such as low-interest financing. Just be sure to look at the total price of the vehicle, the "bottom line," and make sure you get what you deserve.

chasing from a dealer is to buy directly from the previous owner. The obvious advantages? No middleman, and the person selling the vehicle should have an intimate knowledge of it and be able to answer any questions or concerns. He or she may be able to provide you with the entire repair and maintenance history for it. In fact, you should become a bit suspicious if the history, or at least a reasonable collection of receipts, is not available.

When negotiating the price, do consider that you will not be getting any warranty from the owner, which involves risks that you might not be taking if you bought the car from a dealer. You might just mention this fact during the negotiations. It may help bring the price down. Actually stating that you are concerned about having no warranty may also help you discover why the previous owner is selling the vehicle. Especially if the car seems to be in exceptionally good condition, or is a very late model, knowing why he or she is selling it may help put you more at ease in a situation that could, and should, arouse your suspicion. Even then, proceed with caution. A deal that seems too good to be true probably is!

Don't even think of buying a used car unless the owner will allow you or a repair shop you trust to perform a thorough bumper-to-bumper inspection. In most cases, such an inspection will cost less than $100 and it is money well spent. If the tags have already been removed from the car, write the sales contract to include conditions such as satisfactory inspection results, and specify what happens (i.e., the previous owner pays for repairs or the sale is canceled) if the car doesn't pass.

Take these inspection and protection steps for every used vehicle purchase, even for a classic or specialty car, unless the deal is so good that you are willing to assume the risk of an expensive unforeseen problem that makes itself known shortly after the purchase. It is true that the good deals go quickly, but you must resist the temptation to buy in haste. As Roy advises: "If you miss this one, there will always be another pretty car that meets your needs. Be prepared to walk away any time you feel the least bit suspicious or uneasy!"

Buy from a Friend?

Buying a used car from a friend may seem ideal. After all, you will know the previous owner and have a good sense of what kind of care and repairs the vehicle has had. You may even know if it has ever been involved in an accident. But there are pitfalls to this scenario, too.

Since the owner is a friend, you will be more likely to accept their asking price rather than negotiate the best deal and risk hurting their feelings. In addition, you and your friend may take it personally if the car is fraught with problems after the sale, something that probably would not happen with a stranger. If your friend offers to finance the purchase, the gates are wide open for disputes. Even if you believe that it could never happen to you, you must realize that the courts are full of lawsuits between friends over borrowed money. A business deal that seemed like a great idea at the time has destroyed many a friendship.

Buying from a Dealer

Someday, we may be purchasing all our cars over the Internet. However, the vast majority of deals for both new and used cars are still made, or at least completed, on the sales floor of a dealership. We can provide lots of help here, because we have both bought lots of cars over the years. Roy has also spent a number of years working for new car dealers, just a few feet from the showroom, so we are in a position to give you some inside dope. Roy says, "When you are exposed to sales professionals and their most recent customers every day, a little knowledge just has to rub off." He also admits to having "more than a few" social outings with his colleagues from the sales and service departments, during which he picked up numerous nuggets of information about how sales are made. Let's discuss the games of the sales floor in more detail.

The Right Car at the Right Price? Here's What You're Up Against!

You are probably at more of a disadvantage when buying or leasing a car than in any other business transaction. Make no mistake, there are strong emotions involved when most of us look at a car or truck that we *really* like. It is all too easy to be seduced by the feel, smell, and power of a new car. As soon as logic gives way to emotion, you have lost your edge, and the odds of getting "the right car at the right price" are sinking fast! Once the sales person picks up on your "I gotta have it" feeling—and salespeople are keen judges of human emotion—you are playing a game against a stacked deck. You may find yourself wanting to drive that car home today, and you might be willing to jump in and make the commitment to buy right now—to your definite disadvantage!

Roy admits to having quite emotional reactions to cars. Davidson, himself, claims that he hasn't experienced such feelings about cars since adolescence. However, he doesn't like to be reminded of that old Infiniti J30 he bought in a weak moment. (Premium fuel, no trunk space, poor gas mileage. What *was* he thinking?)

To even the odds, you must avoid emotions, or at least avoid showing them, during the entire negotiating process, and *always be prepared to walk away!* You simply must try your best to avoid falling in love with a particular vehicle. Remember that salespersons are trained professionals. It is their job to have you buy a vehicle from them, at their dealership, and for the highest possible profit. To that end, they are very sensitive to any emotions and information they can gather from you.

Salespersons know that, if you do not make a purchase commitment with them today, the chances that you will come back to that dealership to buy your car go down with every hour that passes. They will take up as much of your time as they can, and keep you at their dealership for as long as possible. There are two reasons for this. First, the longer you stay in this showroom, the less time you will have to shop for other deals. (Remember that tiny, precious block of discretionary time?) Second, the longer you stay and talk, the more useful information they can gather which they can use to persuade you to buy that car today and/or direct you to another vehicle that you might fall for even more deeply.

Showroom Floor Tactics to Watch Out For

There is certain crucial information that the salesperson will use to his or her advantage when you visit the sales floor. To begin with, your grooming, clothing, body language, the way you speak, and the car you are driving now will provide an instant and usually accurate assessment of what you can afford and the type of car in which you will likely be interested. The variables that will affect your buying decision are also critical nuggets of information. These include such things as:

- Why you are interested in a particular model
- What you think the purchase price should be
- What monthly payment you can afford
- Your credit rating
- The value of your trade-in and its payoff balance
- How you will use your car or truck (Business? How many miles per year?)
- How much down payment you can make
- Whether the deal will be a cash sale or financed
- What brought you to this dealership
- Whether you are a shopper or are ready to buy (perhaps the most important of all!)

As you begin conversing, the salesperson will try to gather as much of the above information as quickly as possible, but will likely do so in a very indirect way. You will also find that any preferences you express, such as color choices, will influence the cars you will be shown. For example, if you indicate through your conversation that you need the space that a minivan offers and that you love bright red cars, you can bet that you will be shown any large, bright red vehicle on the lot!

You will find that the sales professional will never use anything but the broadest of terms in discussing price, and that no negotiations will take place until you have indicated a particular car or truck that you will accept, whether or not it is in stock at the moment. This means that you have essentially agreed to a particular model, with a particular list of options. The sales professional must have that information, along with your credit qualifications and when you want to take delivery, in order to begin negotiating. This process may take only a few minutes or it can last several hours, but eventually you will be placed in the position of having found a car that you will commit to buying "if the price is agreeable to you."

Let the Fun and Games Begin! (And Guess Who's It?)

When you have found a car or truck you are willing to accept, you should remember a primary rule of negotiations: "The person who wins in a negotiation is the person who is in control." Whoever allows any element of control to be lost, loses! For example, if you surrender your credit card to the salesperson, you have just lost the ability to get up and walk out of the showroom. That is likely to be the real reason they asked for it, not to run a check of your credit.

This is one example of the games that are played every day on the sales floor. We are certainly not accusing all auto salespersons of being dishonest, but the fact is that there are a number of techniques and tactics that can be, and are, used to maximize profits. Here are a few of the most common ones.

Good Guy/Bad Guy

One of the all-time classics! The salesman is always the good guy; it's the sales manager or the person who appraised your trade-in that is the bad guy! After you have spent an hour or so and finally agreed on a price and terms, the salesman "tries to get the deal approved," and may even ask for a deposit or credit card as a show of good faith (which you should resist). He then comes back to say that "they" cannot let it go quite that cheaply. Of course, he will never stop telling you what a good deal you are getting! If you continue to resist, the salesman may impose a sense of urgency, even going so far as to tell you that another salesman has a deal working on the same car. This might be true, but is not likely unless you have chosen a "hot," fast-moving new model. (You won't be getting a great low price on one of those anyway!) If the salesman is still not successful, the sales manager may arrive to take over the negotiation. Actually, that is what you want, because you are now negotiating with the person who has the authority to sign the deal. A good negotiator will not waste his or her time talking to someone who cannot close the deal personally.

Figuring Downward

The salesman will try to express any additional charges in the lowest possible manner. In other words, an additional charge that will raise your monthly payment by $10, a total of $600 over sixty months of payments, will be expressed as "only 3 cents a day" during negotiation. When numbers are expressed like this, the buyer tends to be unaware of how much he or she is actually paying for some sort of extra.

Dealer Installed "Soft" Items

These items, often referred to as "fluff," usually consist of high-profit items that add little, if anything, to the value of the vehicle, new or at resale. In short, they are items you probably don't need or want. You certainly don't want to pay interest on them over the life of your auto loan. If you want an air deflector for your sunroof, you will be better off buying it from a specialty shop than paying interest on it for forty-eight months or more. You may also find that a $200 item costs only $40 over the counter at an auto parts store down the street! Some typical dealer-installed items that you may have to do battle *not* to pay for are:

- Alarm systems
- Body side moldings or wheel well trim
- Door edge guards
- Floor mats
- Mud flaps
- Custom painting, such as pinstripes or gold-look emblems
- Appearance protection items, such as rust proofing, fabric protection, paint sealer, and undercoating
- Window tinting
- Custom wheels/tires
- Locking lug nuts or wheel covers

If the car or truck you are buying already has "fluff" installed, you may not be able to get everything for free. The salesman may actually be more willing to deduct an equal dollar amount by "giving" you some low-profit factory-installed options.

Trade-In Allowance

If you are trading in your old car, you will be shown an "allowance" for it in the sales contract. Watch out for this one! The "allowance" shown will almost certainly be deducted from the inflated "gaff sticker" price, an industry term that means the dealer's mark-up price that includes all of the "tacked on" fees and dealer-installed options. (A gaff, by the way, is a sharp hook used by fishermen to get a large unruly fish into the boat! Get the picture?) Close inspection of the figures will reveal that the dealer is actually paying a lower price for the vehicle. If you add up the trade allowance and the discounts, then subtract them from the sticker price, you will probably see a different purchase price for the new car than the one you just agreed to!

The only sure way to know what your trade-in is worth is to do your homework ahead of time. Buy a *N.A.D.A. Official Used Car Guide* or go to an automotive Web site that lists *Kelley Blue Book* prices. Be honest about the condition of your car, as the person who appraises your car at the dealership will notice any deficiencies. Generally speaking, the "average wholesale" value of a used car that is three to six years old is about two-thirds of its retail value in the Blue Book (even though the Blue Book itself may indicate a higher wholesale price). Most dealers pay an actual cash value for trade-ins that is more like "rough wholesale," generally about 50 to 55 percent of the retail value.

Make no mistake, trading in a car rather than selling it yourself saves you hassles and headaches, and also reduces your risks—but it costs you money! The only exception is trading in a very old or very rough vehicle that isn't worth much money at retail. It will not make much difference whether you sell old Bessie or trade her in. The dealer isn't going to resell this car, and it won't bring much from a wholesaler, either. Your true trade-in value is zero. Any "allowance" given is actually a discount on the new car!

The Finance and Insurance Office

Most new car buyers breathe a sigh of relief when they are finished with the salesperson and are taken to the finance and insurance office (F&I) to "complete the paperwork." They may also be anxious to get the whole thing over with and get on with their busy day. What most consumers do not know is that, with few exceptions, the finance and insurance office is staffed by a person who is 100 percent commissioned and who may be the most skilled and highly paid member of the sales department! The words to always keep in mind when bringing a deal to completion in the F&I office are: The purchase price of the vehicle is NOT necessarily the total you will pay! This is where most "bumps" (add-ons) are made to the original, signed deal. Remember that any bumps will add to the total amount you are financing and you will pay interest on them over the life of your loan. In other words, you will be paying more than the cost of the add-ons! A deep discounted purchase price may end up being a high-profit deal if many high-profit add-ons are included in the total cost. F&I is often where you are introduced to the "soft" add-on items discussed earlier, or one or more of the following.

Extended Warranties

These take three forms: Extended manufacturer's warranties; extended warranty coverage that is good only at the selling dealer (actually a service contract); and mechanical breakdown insurance. You should approach these in the same way you would consider an insurance policy. Ask for a copy of the actual policy, not just the sales brochure. Read all of the fine print carefully, then read it again. Note any questions or concerns and read it a third time. The rule is, *if the warranty does not specifically say that something is covered, it is not covered!*

Be sure to understand the cancellation terms and rights you have. Ask if you can sign up for the coverage later. Generally, extended warranties do not have to be purchased at the time the car is bought. Take the policy home and take the time to read it, or ask someone to help you review the fine print. Compare the coverage of the extended plan to the new-car warranty to see just how much additional benefits you are really buying. Do not sign or agree to accept anything you are the least bit confused about!

Trade-in Payoff Games

Sometimes, the payoff for your trade-in is shown higher than it really is. You may be asked how many monthly payments you have remaining, and "payoff" may be shown as the sum of those payments. In reality, the payoff amount is the sum of the unpaid balance and any early payment charges assessed by the bank. There should be no interest, and no processing fees assessed by the dealer. In order to "expedite the paperwork," you may be given an estimated figure for payoff in the F&I office, and asked to sign a blank payoff authorization form. Do not agree to this! Once you have signed that blank form, you have agreed to whatever is filled in later, regardless of your actual payoff amount. You might as well hand over a signed blank check! The best way to protect yourself from this game is to simply call the bank and request payoff information yourself, before you go to the dealership. Now you know what the amount on that authorization form should be.

Interest Rate Bump

There are two variations of the interest bump. In the first, you may be told that you cannot be approved by a certain lender at the low-interest advertised rate because of your credit rating or some other reason, but you can get approval from another lender at a higher interest rate. The second variation is that the rate shown on the final sales contract is different than that which the sales person quoted you. Since most consumers are not capable of computing compound interest payments while sitting in the F&I office, you may not realize this is happening. Once again, read the fine print and all of the filled-in numbers very carefully. Question anything you are unsure of and do not sign anything until you are comfortable. Remember that interest rates are not non-negotiable. If you have arranged your own financing and come to the showroom "bought at the bank," this bump simply cannot happen.

Lower Payment with Longer Term

This game is often played when a buyer is uncomfortable with a monthly payment above a certain dollar figure. The F&I person simply extends the term of the loan to reduce the payment. In the process, the interest rate will likely increase along with the term of the loan. You will probably find that a payment reduction of only $20 a month will cost you an additional $2,000 or more in the total of your payments. Once again, get the numbers ahead of time by arranging your own financing and a surprise like this will not affect you.

Yes, You Can Win!

You can beat them at many of these games by—you guessed it!—arranging your own financing before you ever walk into the dealership in the first place, and coming into the sales office with a particular model,

What the Factory Invoice Really Means

Many times, cars are advertised for sale "$100 over invoice," or even "under invoice." Are the dealers giving the cars away? Don't you believe it! Theoretically, the factory invoice is the price that the dealer paid for the car. That would be true if the dealer only bought that one vehicle from the manufacturer. There are a number of things that cause the dealer's actual cost to be lower than invoice. For example, the factory will allow discounts for the quantity of vehicles ordered and/or sold by that dealer during the year, or provide incentives such as certain option packages that are supplied at no charge on specific models.

Sometimes, the manufacturer will discount less-popular models, just as a retailer will discount the price of non-moving inventory. The dealer actually may not know how much discount they will receive on a particular car in advance, because the incentives and discounts are often figured and paid by the manufacturer at year-end. This is why the ads promising factory invoice pricing tend to occur near, or after, the end of a model-year. However, the experienced sales professionals at the dealership will have a pretty fair idea of just how much room they will have below the invoice. In most cases, a car that is not specially discounted by the manufacturer will be very hard to buy at or below invoice, but it is realistic to expect a deal that is $500 to $1,000 over invoice without the fluff of dealer's fees and tacked-on items. If you are paying for the fluff, then you should expect to get very close to invoice.

option list, and color choice already established. By doing this, you are already telling the salesman that you know what you want and that you are shopping for price.

Approach car buying like a job. Come equipped with notebooks, a folder or large envelope for brochures, and a calculator that you know how to use. Review the math involved in figuring out percents and interest ahead of time. Be ready to quote numbers and options, with your

Internet research results standing behind you.

If the model you have selected is a popular one that should be readily available, you are also sending a signal that you have probably shopped other dealerships already. At this point, the sales professional will simply check the inventory and the negotiations can begin. Before you enter into the negotiating process, take one last opportunity to review the options you want, and decide exactly what you will and will not accept in the way of alternatives (in seating, engine, accessories, color, and everything else).

Do this at the first dealership you visit, so you know exactly what is available as you begin comparison shopping. If you have received quotes on the Internet or by phone, there is another reason to review the options. You may find that all of the option packages and choices for a particular model are not listed at that dealership, or that the exact options you requested are not available on the vehicle in stock. This is often done in order to simplify the quoting process and encourage more dealers to participate. Unfortunately, this process may also give the sales professional an opportunity to bump up your options, or even bump you up to a higher-profit model when you actually arrive at the showroom.

One last thing to do before you begin the negotiating process is to take a test-drive *in the exact vehicle you wish to buy.* Go for a drive that is long enough to allow you to objectively evaluate the ride, handling, idle quality, noises, and the convenience of things you will be using regularly. Is the wiper switch hard to find? Do you have to take your eyes off the road and look downward to change the radio station? Is the color and brightness of the instrument cluster lighting pleasant or annoying? Can the driver's seat be positioned so that you will still be comfortable after a four-hour drive? If you are a smoker, are the ashtray and cigar/cigarette lighter easy to see and reach, even after dark? Now is the time to get acquainted intimately with this machine that will be such an important part of your day-to-day life for quite some time. Do be reasonable in how many miles you rack up on the test-drive, a new car will be somewhat devalued by fifty or so miles on the odometer if you decide not to buy it.

By the way, if you encounter a "new" car with several hundred miles on the odometer, that may be an opportunity for you to demand a discount!

Before your test-drive, please pay particular attention to the Test-Drive Checklist (see fig. 2.3). Read it through and copy both pages to take with you when you test drive the car. Go right through the list, item after item, and take your time doing it.

When everything is acceptable and you will agree with the salesperson that they have you in "the right car," it's time to get down to the business of price, terms, and delivery.

Generally speaking, you will be able to negotiate a better deal on a vehicle in stock, rather than one that has to be ordered or obtained from another dealer. The reason is simple: there are additional costs and risks for the dealer in ordering or exchanging a vehicle. One reason that Internet quotes generally provide favorable pricing and a time limit for the price guarantee, if any, is that they are based on in-stock vehicles.

You should also be able to obtain, or at least look at, a copy of the factory invoice for a vehicle that is already in stock. In obtaining the invoice price, you can negotiate dealer profit, rather than purchase price. This puts you in a very good position to haggle. If the dealer will not share the invoice price with you, be very cautious as you proceed with the deal, or walk away from it altogether. This is a sure sign that you have encountered a very sharp, highly trained sales staff, and that you have given the salesperson some indicators that will make him or her begin negotiating "full boat" at the manufacturer's suggested retail price (MSRP), called the "sticker price." Even worse, they might try you on the dealer's mark-up price that includes all of the tacked on fees and dealer-installed options, commonly referred to within the industry as the "gaff sticker price" that we mentioned earlier.

Up to this point, we have tried to provide guidance in getting the most value and avoiding the pitfalls (well, most of them, anyway) when picking the right car and taking possession. Beginning with chapter three, we will shift gears and begin talking about the rest of your automotive ownership experience with this Perfect Car (or truck) of yours. We'll discuss all kinds of useful stuff, and have some fun, too. So, please come with us now, into the realm of care and repair.

Test Drive Checklist

Powertrain

Engine power

Should seem adequate for your style of driving. Listen for engine noise. In some cars there is more noise transmitted into the passenger compartment than others. Make sure the amount of engine noise and vibration you are experiencing is acceptable to you, because it probably isn't going to get any better.

Smoothness

Smoothness of acceleration and deceleration? Make sure that you have a secure feeling that the car is going to move appropriately when you press down on the accelerator.

Idle quality, cold and warm engine.

Listen to it when you just start up, and just before you shut off the engine. The idle should be acceptably smooth and free from vibration both before and after driving.

Ease of starting, cold and warm engine.

Admittedly hard to check on one test drive. Still, pay attention to the sound of the car cranking (check Chapter Four for definition of cranking). It should start immediately after one or two cranks. If you have to key it twice to start it, forget it.

Ease of checking/filling fluids under the hood

Look under the hood and locate all the dipsticks and filler covers. Note the position of the battery and the spark plugs.

Automatic Transmission

• Smooth engagement?
• No severe jerks when you put the car from neutral into drive?
• Smooth, sure upshifts and downshifts?
• No feel of slippage?
• No whine or grinding noises?

Figure 2.3—Test Drive Checklist

Manual transmission

- Smooth clutch engagement?
- No grabbing or chatter as you engage the clutch?
- Normal clutch pedal feel and free travel?
- No clutch slippage? If the clutch is slipping the engine speeds up but the car does not. Especially in the higher gears.
- Smooth, easy shifting?
- No whine or grinding noises, all gears?

Front wheel drive

- No noises (clicking or clunking) in tight turns? Could mean bad CV joints on a used car or a defect in the drive axles on a new car.

Rear wheel drive

- No gear whine or rumbling noise, all speeds? Noise could mean wear or damage on the rear (drive axle). In the old days they used to pack differential lubricant with sawdust to cover up those noises.

Steering

- No pull to one side?
- No wander?
- Ease of steering when parking? Try parking in several different kinds of spaces. Pay particular attention to the turning radius, which is very important in parallel parking. Both Roy and Davidson had old Dodge Minivans, which were very disappointing in terms of turning radius. Tight radius is good. Too large a turning radius is not good.

Ride/Handling

Does the car have an acceptable ride? This is a subjective thing. Some cars, particularly sports-type cars, ride stiffer than others. Try out the car on various road surfaces if possible. Make sure the ride is acceptable to you. Make sure you have complete control on bumps and turns with no pitching and rolling, or excessive leaning, in curves.

Brakes

Test the brakes on a stretch of empty road. Aim car straight ahead, speed at about 35 mph. Remove hands from wheel and step hard on the brakes. Car should come to a fast rolling stop (if you have ABS!) with no fishtailing or pull to either side.

Figure 2.3—Test Drive Checklist (continued)

- Do the brakes have a firm, high pedal feel?
- A sure smooth brake application, no spongy feeling in pedal?
- Stopping power adequate?
- Parking brake effective and easy to apply/release?
- Equipped with anti-lock brakes? Is it a 2 or 4 wheel system? That's a question you should be able to get off the sticker.

Driver Conveniences

First, pay particular attention to both driver's and passengers' seating. Take time to adjust the seats so they are comfortable for you. Adjust distance from dashboard, height, and inclination. When you think you are perfectly comfortable in the seat, adjust the steering wheel. Get in touch with how your upper back, lumbar, and legs feel, and try to guess how they will feel after a, say, 600 mile drive in that car. Some cars are just not right for some people. Roy, for example, cannot get comfortable in a Corvette, because he is too tall. He has to scrunch down to see through the windshield. If the car you are trying out is not for you, now is the time to find out.

Make sure the audio system sounds good to you. Play with the knobs, try out different stations, and bring a CD to play in the CD player if it has one. Note the position of the dome light and try out the reading lights, and make sure switching on the reading lights is easy and convenient. Open the glove box and any other compartments.

Try the accessories and see if you are comfortable with them. If it is going to distract you every time you turn on the windshield wipers, open and close the windows, adjust the radio volume, etc., then try another car. We mean it. Driver comfort is going to be a big part of whether you are going to be happy with this vehicle or not.

Controls/accessories convenient and easy to use:

- Wipers
- Audio system
- Lighter/power outlets
- Dome and reading lights.
- Head Lights
- Instrument lights color/brightness
- Windows
- Locks
- Cruise control

Figure 2.3—Test Drive Checklist

- Navigation system
- Seat/steering wheel adjustments
- Cup holders

Safety Features

Ease of installing/accessing child restraint seats?
How many airbag restraints and where they are located?
Is airbag disabling switch available for this model, if needed?
Are the headrest restraints comfortable?
Are the jack and spare tire easily accessible and easy to use and mount?

Fit and finish

Exterior

Are the body panels, trim, doors, and windows correctly aligned and working properly? Take a good look at the body. Make sure everything fits like it should; for example, all moldings and stripes should be on straight. Look for evidence of undisclosed repainting or major repairs. Repainting is not necessarily bad. But if this is a brand new car and you see evidence of overspray or masking marks around the edges of the body panels, it's a sign that major repairs may have been done to your "new" car.

Interior

Look at the condition of upholstery and trim. In a used car look for bad stains, torn areas, or cracked leather on both seats and side panels. Wear of pedal pads should be consistent with odometer reading. Check this out. At 20,000 miles on the odometer, the pedal pad should not be worn out!

Figure 2.3—Test Drive Checklist (continued) This checklist applies to both new and used cars. Remember that the final inspection and test drive are your last opportunities to decide whether this is the car for you. So pay attention! Even the little details can wreck your relationship with your car. So copy this list and take it with you when you test drive. Oh, and don't be afraid of being a pain in the neck!

CHAPTER 3

GETTING OFF TO A GOOD START!

Buying the Right Gas, Oil, Battery, Fluids, and Tires

It's always a happy moment when your new (or new to you, anyway) car is finally in the driveway. If you have followed our suggestions from the previous chapter, chances are you now have a car with real possibilities for providing miles—and years—of good dependable service. And it's perfectly okay to feel a little pride in your new purchase. You went and did the math, and asked yourself some hard questions about your automotive needs. Armed with the results, you then went out and did the legwork, and the car in the driveway is your result.

After the purchase it is all—or nearly all—maintenance. What you do now in terms of care and upkeep will be a deciding factor in how long your car keeps running. It could be a really great car, but it will age prematurely if you don't take care of it. For starters, let's take a close look at your car's fuel and fluids.

Oil companies and other automotive corporations try to make us believe that each of their products is better than the others on the market. Common sense, however, tells us that every company's

Figure 3.1—This '93 Chevrolet Camaro has had an interesting history. It was first titled as a leased car in Florida, then found its way to Kansas, then ended up back in Central Florida again. It still looks great at 138,000 miles, and the owner says it "runs like a rocket!"

product cannot possibly be the best. The differences between the various brands of gasoline, oil, and other fluids your car needs lie in the specific additives that each company uses. The additives all largely accomplish the same benefits, but they do it using different blends and formulas. These performance additives are patented compounds, which can only be used by the companies that own the patents. So each company must develop its own patented formulas for their products. This is the main reason why different brands of fuels and oils differ from each other.

Is any one brand of gas or oil better for your particular car? Well, not a whole lot better! We'll tell you how you can, within limits, find the optimum fuel for your car later in the chapter. Meanwhile, how can you make sure that you are "feeding" your car the correct fuel and fluids? Step one is to open the glove box. Then, take out the owner's manual and maintenance schedules, and begin reading.

We don't believe that everything car manufacturers and oil companies do is perfect. Neither do we believe every word they say. They are corporations, and just like other corporations they tend to base important decisions on corporate concerns, not necessarily those of the general public. The people who actually write the manuals, however, are technical writers, engineers, chemists, road testers, technicians, and other professionals. It is part of the ethos of technical people to "get it right" when confronted with a technical question. So we trust that they go to a great deal of trouble and expense to ascertain which type of engine oil, power steering fluid, and transmission fluid will provide the best performance and protection for your specific make and model. Sometimes they make mistakes, but they go to great lengths to determine the life expectancy of those fluids, as well as a host of parts that require periodic replacement, under many different types of driving conditions. This is the information upon which their recommended service intervals found in the maintenance schedule are based. No one knows your car's requirements better than the people that designed, built, and tested it!

In the owner's manual, you will find several pages devoted to the manufacturer's recommendations for the exact type and grade of every fluid and chemical your car needs—even upholstery and carpet cleaners are covered! All you have to do is make sure that the fluids you use meet the standards they specify. This does not mean that you have to buy all your fluids from

USED CAR OWNER'S MANUALS

Owner's manuals can be had for used cars if they (the cars!) are not too old. Usually, you can obtain ordering information from the dealer's parts department or by calling the local customer service number for the manufacturer. In addition to the manuals, there are a lot of other sources for recommended maintenance. You can buy some books at the local auto parts or discount store. There are even some computer software programs that tell you when maintenance is needed, then give you a neat little tracking system to record service and repairs as you go.

the dealer, or that you must have all of your maintenance done there. The container that any good quality fluid comes in is clearly labeled to indicate the standards it meets. If it isn't, don't buy it for your car! For example, if your car requires an engine oil rated by the American Petroleum Institute (API) for service class SJ, and a viscosity rating of 5W-30, simply look for a container of oil that has those ratings on the label. We will talk about what the fluid ratings mean a little later.

OCTANE AND FUEL ECONOMY

When the engine knocks from a low-octane fuel, the computer retards the ignition timing to make the knock go away. This results in a slight loss of power and fuel economy. Usually it's not noticeable, especially since most late models can actually tell which cylinders are knocking and retard the spark on only those cylinders. Theoretically though, a car that is knocking on 87 octane might get slightly better economy on 91 octane fuel due to that effect. However, if a car rated for 87 octane runs poorly on it, something is wrong and should be corrected. A light knock won't hurt anything and will likely as not be corrected by the car's computer.

Thoughts for Fuel

Automotive fuel is a blend of refinery product plus several additive compounds designed to adjust octane, reduce deposits in the engine, help the fuel burn cleaner, and "seasonalize" it (a Roy-ism) to burn more efficiently in hot—or cold—weather.

The primary decision you need to make concerning fuel is whether or not it has the right octane rating for your car. Are all gasolines the same? No, for the reasons we mentioned earlier in the chapter. However, all octane ratings are measured the same way, and you should use the rating recommended in your owner's manual. The vast majority of late model cars and trucks are designed to fun on "regular" grade fuel, which has an octane rating of 87. There are a few exceptions, where the recommended fuel will be "mid-grade," with an octane rating of 89, or even "premium," a rating of 91.

Octane rating is a measure of how fast the gasoline "burns" (expands) inside your engine's cylinders after it is ignited. This is controlled by "anti-knock compounds," used to slow down the burning of the gasoline inside the engine. The higher the octane rating, the slower the fuel burns. The slower the burn, the cooler the engine's internal temperature remains, so

THE OPTIMUM FUEL FOR YOUR CAR?

Yes, there really is a difference in the specific chemical formulas used for fuel additives by the various oil companies. There is no rule of thumb, however, or any way that we know of to tell which car really runs better on which gas, including the cheap brands. If the oil companies are to be believed, they are always improving the additive mixes, anyway!

The very slight differences in the individual engines, even of the same make and model year, the driving habits of the driver, and subtle changes in driving conditions, make finding the optimum brands of fuel and oil for your car very hard to do a priori. It is best to rely on your own observations and experience with your own car. Roy's father-in-law always swore that his Chrysler got the best gas mileage on Chevron and the worst on Exxon, even on cross-country trips from North Carolina to California. Now, he drives a Ford Crown Vic, and he uses nothing but Exxon!

Davidson's dad was a Gulf Oil customer. No matter what car he was driving, the "big orange sign" was his favorite stop. Davidson himself usually buys Texaco, but that's because they sponsored Milton Berle's TV show and the Saturday afternoon live opera broadcast. Go figure!

the engine is less apt to knock or "ping" on acceleration. If the engine is running at a speed of 3,000 revolutions per minute (rpm), each cylinder is burning a charge of gasoline 1,500 times per minute, so we are talking about a very small fraction of a second between the burning time of 87, 89, and 91 octane gasolines. That is the only difference octane rating makes!

If 87 octane is recommended for your engine and the car runs on it without pinging, there is no significant or additional value to be obtained by using a higher octane fuel. If the engine does knock or ping on acceleration using the recommended fuel, the problem should be diagnosed and corrected. You should not have to use fuel above the recommended octane to achieve optimum performance.

Now that you are using the right octane fuel for your car, it is entirely possible that a particular combination of engine, driver, and accessories will perform better on one fuel blend mix than another. So you should try several different brands of fuel to see if your car likes one better than the rest. Check your fuel economy, the general feel of the car's performance, and the ease of starting with the engine both cold and hot to judge whether or not there is a difference. Be sure that you conduct these tests under routine conditions and be sure to try at least three or four tankfuls of each brand, because the difference may not show up right away. Also, since the additives are sometimes blended into the fuel after it leaves the refinery (at the storage terminal, or even as it is loaded into the tanker truck), don't be surprised if there is a real difference from one tankful of gas to the next in the same brand of fuel!

Motor Oil: Your Engine's Lifeblood

We refer to motor oil as the lifeblood of your car's engine with good reason—your car's engine will die a nasty death without it! Motor oil is similar to the blood in the human body in several ways. It is circulated throughout the entire engine under pressure and, if the circulation becomes blocked, the engine will self-destruct. Also, if it becomes too thin, too thick, or contaminated, it cannot do its job properly. For these reasons, changing your oil as recommended in your vehicle's maintenance schedule is an absolute must.

Motor oil also provides about 40 percent of an engine's cooling needs. There are a lot of components in the engine; some of the warmest ones, such as the pistons, rods, crankshaft, and main bearing assemblies, have no direct contact with the cooling system passages in the engine. These parts are cooled as the oil carries away the heat; the oil itself is cooled by the airflow over the oil pan at the base of the engine and/or by passing through parts of the engine that are exposed to the cooling passages. Some engines also have an actual oil cooler, via which the oil can be exposed to airflow or to the cooling flows offered by passage through the radiator.

MYTHS ABOUT SYNTHETIC OILS

Myth: Synthetic oils have only been on the market for a few years.

Truth: Synthetic oils were developed in the 1930s. To this day, they have still not received wide popularity.

Myth: Using synthetic oil will cause oil leaks.

Truth: Since synthetic oils flow better, they may seem to make an already existing leak larger. They also provide better "wetting" of surfaces, so a small spot of synthetic oil on an engine component may form a larger stain area. There is no evidence that any gaskets or seals are damaged by them.

Myth: Synthetic oils are not compatible and will cause engine damage if they are mixed with each other, or with conventional oil.

Truth: Although some of the earliest synthetics would turn to a jelly-like substance if they were mixed with conventional oil (we're talking the 1930s here!), there is very strong evidence that modern oils can be mixed without disastrous results. However, Davidson and Roy do not recommend mixing brands of any fluids, because there is a risk of upsetting the carefully designed balance of the additive package. Why take the chance?

In your owner's manual—and displayed on the oil filler cap of most late model engines—you will find the manufacturer's motor oil specifications. The two ratings to concern yourself with are viscosity and service classification. Both are the result of standardized testing and measurement procedures established by the American Petroleum Institute (API).

Viscosity is a measure of the oil's thickness and also indicates the temperature range in which it will perform well. It is determined by measuring the oil's flow rate at specific temperatures, including some quite hot. It is rated in graduations of five and ten, up through 50. The higher the number, the higher the temperatures in which the oil is designed to work. If a *W* appears following a viscosity rating, it indicates that the oil can perform over a somewhat wider (hence the *W*) temperature range than one without the *W*. For quite a few years, virtually all auto makers have recommended multi-viscosity oils, with ratings such as 5W-30 or 10W-40. These have the properties of a thinner, low-viscosity oil when cold, and still provide the protective lubricating qualities of a thicker oil at high temperatures. This combination allows easy starting and warm-up for a cold engine, as well as superb lubrication for today's engines, which are designed to run at higher temperatures. These oils even improve fuel economy slightly and help today's engines reduce exhaust emissions. That is why you will often see the words "energy conserving" on the container label.

The API service classification indicates the oil's ability to provide adequate cooling and lubricating performance under severe driving conditions. The API service ratings are a graduated series of two letters. Those that begin with *S* are for "service class," for gasoline engines, and *C* for "commercial class," for diesel engines.

In the service class for gasoline engines, the *S* is followed by a letter from *A* to *J*, the *J* indicating oil designated for the most severe use. In fact, all service class categories other than *SJ* are currently obsolete. The reason is that new engines, made to ever-closer tolerances and more demanding specifications, require ever-better motor oils. New classes of motor oil keep coming: service class *SL* may be just around the corner!

You may be able to find older designation motor oils on some gas station or auto parts store shelves, but they should only be used in the less demanding, earlier model engines for which they were approved. It is perfectly all right to run *SJ* oil in an engine that only requires *SG*, but you would be flirting with disaster by running *SG* oil in an engine requiring *SJ*. Taking that chance could result in a quick and painful end to your relationship with your perfect car! It's probably best to shun all earlier rated oils and only use *SJ*, or whatever higher rating becomes current as time goes on.

A Word About Synthetic Oils and Aftermarket Additives

With all of this talk about high temperatures and demanding engine conditions, you may wonder about using synthetic motor oils or after-

WHOSE MAINTENANCE SCHEDULE?

When deciding when to have your fluids changed, your hoses and belts replaced, your brakes inspected, and all of those other adjustments and lubrications, note that there are often three maintenance schedules to choose from: the manufacturer's "normal service" schedule, their "severe service" schedule, and the dealer's recommended schedule.

An honest assessment of your driving habits and conditions can tell you whether or not you fall into the "severe service" category. For example, the maintenance schedule almost always defines "towing a trailer" as severe service. If you tow a trailer several times a month, you would consider that severe service. However, towing a trailer once or twice a year might mean that one oil or transmission fluid change should be done sooner, but it would definitely not be classified as severe service.

What if the dealer provides a third schedule? Is that one better? Compare it to the manufacturer's schedule. You will likely find that it is similar to the severe service schedule in most respects, and may even exceed it. A closer look may also reveal a long list of "check this" and "inspect that" for items not even mentioned in the manufacturer's schedule. It will certainly do no harm, but you may be spending your maintenance dollars on service that isn't needed, or paying for a long list of inspections that you could do yourself. Roy refers to this as "Christmas presents you give all year long"—give to the dealer, that is!

market additives that claim to extend oil change intervals and/or provide increased protection under severe driving conditions.

As with fuel recommendations, the manufacturer's motor oil recommendation is carefully arrived at after exhaustive laboratory, proving ground, racing, and actual highway testing by a large team of professionals. You can rely on their recommended oil viscosity and service rating to provide more than adequate protection for your car as long as you change the oil regularly.

However, synthetic oils do have advantages that could make a difference under really severe conditions, such as racing, extreme temperatures, or pulling heavy loads. These advantages include higher natural viscosity, higher detergent properties, improved protection against friction, no harmful impurities, and better resistance to sludge and varnish. These advantages add up to what Roy calls "better abuse tolerance"!

Synthetics also cost more to produce, and purchase, than conventional petroleum-based oils. Whether the increased cost is justified by the increased performance—beyond the needs of most drivers, in our experience—is a question that can best be answered by carefully and honestly evaluating the severity of your driving demands.

Do not use extended oil change intervals as a basis to justify using synthetics, as nearly all manufacturers will void your warranty if you allow your oil changes, or other required maintenance, to exceed their specified service intervals. Some blended oils, comprised partly of synthetics and partly of conventional, petroleum-based oil, are also on the market, and may provide a compromise between the high cost of full synthetics and the need to change your oil often enough to protect your engine warranty. While full synthetics cost nearly four times as much as high quality conventional oils, the blends generally sell for only 60 to 80 percent more than conventional oils.

Aftermarket oil additives are another matter altogether. Some people swear by them, while others disparage them with names like "snake oil." Many of these products contain the same compounds that are used in motor oils in various levels of concentration. A few contain ingredients not found in conventional or synthetic oils.

These are usually of questionable value in improving performance or protection. Some additives could provide benefits if added to oil that has been left in an engine for so long that its additives have been used up, but mostly they just add more of what is already there. If an additive is too strong, it can actually harm performance.

Remember those professionals who work to carefully formulate the additive package in your oil. When you add new ingredients to that oil, you are likely to upset, for better or worse, the balance of features provided by that formulation. For this reason, you will often see wording included in your owner's manual stating that extra additives are specifically not recommended. Our advice is generally against the use of additives, unless they are actually supplied by the manufacturer.

By the way, since your engine's oil is, indeed, its lifeblood, you should be sure to check it regularly. While your new car should not consume more than about a quart in two thousand miles, it is a good idea to open the hood and check everything—the oil, transmission fluid, battery, drive belts, and hoses—once a week. The oil should be checked first thing, while the engine is still turned off.

Overfilling can cause the oil to foam, in which state it cannot lubricate properly. It is better to have the oil level a bit low than a quart or more overfilled. If you are in doubt about what the markings on the oil dipstick mean, or where the dipstick is located on your car, guess where you should look. You guessed it! Some more good technical training included in the owner's manual at no extra charge! We'll share lots of tips about checking the other items as we go along.

Other Necessary Fluids

Now that we have made it past your car's basic needs of fuel and motor oil, let's take a quick look at those other fluids needed by your car for a long and happy life. The basic formula for finding the right transmission fluid, power steering fluid, drive axle lubricant, engine coolant, and brake fluid is the same as for motor oil: simply look up the required ratings and choose the fluid accordingly.

Brake Fluid

Besides motor oil, the only other fluid that has a synthetic alternative is brake fluid. Synthetic brake fluid is easy to spot, because it carries a rating of DOT 5, instead of the DOT 3 or 4 ratings carried by conventional brake fluids. If your car or truck is equipped with an antilock brake system (ABS) your owner's manual will likely instruct you to "use only DOT 3 brake fluid from a clean, sealed container." Adhere to these instructions carefully. DOT 5 synthetic brake fluids can and do attack certain internal parts of ABS components.

DOT 3 and 4 brake fluids absorb moisture from the air if they are left in an unsealed container. In humid conditions, brake fluid can become contaminated in as little as twenty minutes, causing system damage and

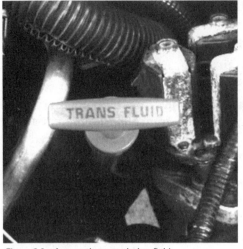
Figure 3.2—Automatic transmission fluid

failure if it is installed in your car's brake system. Whenever possible, the brake fluid level should be checked without removing the cap, by looking through the translucent master cylinder reservoir. It should be changed every two years or fifty thousand miles, whichever comes first, as recommended by the National Highway Traffic Safety Administration (NHTSA). If you must remove the cap to check your fluid, check it infrequently, only once or twice a year, and recap it quickly. Nearly all cars are equipped with a warning light that will illuminate if your brake fluid level becomes low.

Automatic Transmission Fluid

The three common types of automatic transmission fluid currently in use are Dexron III, Mercon V, and Chrysler Type E. Any deviation from the manufacturer's recommendation (found in the owner's manual, of course) will result in at least somewhat reduced performance from the transmission, including such things as rougher shifts and chattering (a high-frequency bumping sensation) on acceleration. The fluid level is almost always checked with a dipstick, much like the motor oil, but in most cases it is checked after the vehicle is warmed up. Check it with the engine running and the transmission gear selector placed in park or neutral, whichever the manufacturer recommends for your particular transmission. Also, checking the fluid level weekly is a good idea.

Other Fluids

The other fluids under the hood that should be inspected weekly are the power steering fluid, engine coolant/antifreeze, and windshield washer fluid. The most important thing here is to use the right type of fluid for your car. Mixing brands doesn't make any difference with these fluids.

HOW BATTERY WARRANTIES WORK

You will often find that there is a significantly longer warranty period for a higher capacity, more expensive battery. However, that longer warranty may not be worth as much as you think. Consider the fact that nearly all battery warranties are prorated, with the value of your battery determined by the number of months the battery has been in service. You get free replacement for only a short period of time, usually a year or less. After that, your allowance toward replacement of the defective battery is reduced every month until the end of the warranty period.

For example, let's say you bought a new battery with a suggested retail price of $60 and a sixty-month warranty. For the first six months, you get free replacement if it fails. But if it should fail after the typical thirty-six months, your replacement allowance would be reduced by 36/60, over 50 percent. So the replacement battery would cost you 36/60 of the current full retail price of $60, or $36.00. If the store has batteries on sale for $34.95, you would actually be paying more for the replacement under warranty than by simply buying a battery on sale.

So, how can you tell when it is time to replace your battery? A battery's life is affected more by age and usage than by miles driven, so you can begin by calculating its age. Despite the fact that many replacement batteries carry warranties of seventy-two months or more, the actual life expectancy of a battery is about half of that. Add in factors such as extreme temperatures or increased electrical power demand, and the life expectancy becomes even shorter. Quite simply, if your battery is more than thirty-six months old, it is living on borrowed time!

In the case of the windshield washer, there is a difference between summer grade and winter grade (which contains antifreeze). Be sure not to get caught in sub-freezing temperatures with a tank full of summer washer fluid, because those windshield washer reservoir tanks can be quite expensive to replace when they break from freezing. Also, there

Battery Warranty Pro-Rate Example
Allowance for a 60-Month Warranty

Months	Percent
6-12	Free
13-18	30%
19-24	40%
25-30	50%
31-36	60%
37-48	80%
49-60	95%

After 31 months from the date of purchase, you would pay 60% of the full, suggested retail for a replacement battery under this warranty.

Figure 3.3—Battery Warranty Table

Figure 3.4—This pretty little '91 Chevrolet Corsica with 179,000 miles on it has only been in the care of its current owner about a year, but it has been beautifully maintained. He insists that is just temporary transportation until he can afford a new car, but he is not working too hard at replacing this reliable ride!

YOU CHANGED THE BATTERY AND NOW THE RADIO DOESN'T WORK!

Darn! Modern car radios are designed to frustrate thieves as well as to sound nice when you're cruising along listening to Garrison Keillor or today's NASCAR race. So, when they are disconnected from the battery, either by being ripped out of the dashboard by brigands or by having your battery changed, they may revert to an inactive mode, requiring the punching of a secret four- or five-letter code into the radio keypad. Some models even require a direct connection to the manufacturer's computer database at your dealer's service department to wake them up! The concept is that a thief will not be able to make your radio play in any other car, so it becomes useless! If you bought your car new, of course, you probably have nothing to worry about. Simply find the little card they gave you with the code on it when you bought the car and punch it in. You did save it, didn't you?

If you bought your car used, though, you might have a more serious problem. Leaf through the owner's manual if you have it; the previous owner might have stuck the code in there. If you can't find it there, take out the ashtray and look on the bottom. If you are lucky, there might be a sticker with the code number on it. If there is no code to be had anywhere, you may have one of the more sophisticated antitheft radios, which are actually smart enough to know whether they have been installed in the right car. Time for a visit to the dealership!

Chances are, though, you will have to pull the radio, write down the serial number of the radio, take it to the dealership of your make of car, convince them of your ownership of the vehicle—and the radio!—and beg them to please request the proper code from the car manufacturer. (Pleeeze!)

Figure 3.5—Battery ratings are found on the labels.

are a few vehicles, especially minivans and SUVs, that have a second washer reservoir in the rear.

While you check the washer fluid, take a moment to look at the condition of your wiper blades. Turn on the windshield washer and notice how well the blades clean the glass. The first sign that your wipers are worn is streaking across the windshield as they wipe. While some streaking can come from accumulated dirt and oils on the glass or blades, continued streaking means it is time to replace the wiper blade inserts. If you let them go too long, the glass can actually be scratched, causing permanent damage. Don't be surprised if you need to replace wiper inserts every six months or so. Their life depends on exposure to the sun and other environmental factors, not on how much they are used.

Battery Checking and Replacement

No component on your car does as much work so quietly as your battery. Unfortunately, when the battery begins to fail, it remains completely quiet. It will die without a single squeak, knock, or groan!

To help you replace your old battery at the right time—after it begins to weaken, but before it fails to start your car—you can look and listen for signs of impending failure. (Roy insists on referring to battery failure as "terminal illness," despite the violent reaction this always brings from his audience...) Begin with a visual inspection of your battery (if it is completely sealed up in a box or buried under the fender somewhere, just skip ahead to "have the battery checked by a professional"). Since batteries produce explosive gasses, wear eye protection. And don't come anywhere near a battery with any source of combustion, such as a lit cigarette.

Look for warped or swollen areas on the battery's case. These are signs that the battery is overheating internally, or that it may have been frozen. A swollen case is always a dire sign. If the battery has removable caps, open them and look at the fluid inside the cells. If the fluid is bubbling, or you can see the tops of the plates exposed above the fluid, it is time to start shopping for a battery. Adding water to the battery, like we used to do in the old days, probably will not help. The way modern batteries are designed it is likely that the fluid evaporated over a long period of time and allowed the dry section of the plates to become damaged, weakening their capacity to produce electrical power.

By the time your battery is old enough to be reaching the end of its life, you should be very familiar with exactly what your car sounds like as it cranks every morning. Whenever the starter seems to be turning slower than usual, the battery is the prime suspect. This is especially true if the engine cranks somewhat slowly after the car has been sitting overnight, but sounds normal when it is started after being turned off for only a short time.

Whether the indicators you notice are visible or audible, to confirm the diagnosis you can have the battery checked by a professional. Many battery dealers and auto parts stores will check your battery, and sometimes the charging system, too, for free. When it is time to replace your battery, make sure that the new battery will provide adequate power and will last at least as long as the old one did. The good news is that getting the right replacement battery for your car can be as easy as going into the local auto parts store and asking for it. There is a book that the salesperson uses to look up your car's make, model, engine size, and electrical accessories. The book lists the recommended battery for your car. If you bought your car new, you can also determine which battery to buy yourself by looking at the original battery.

Battery ratings are found on the labels, generally on the top or sides of the battery case. There are two important ratings: the first is two numbers designating the battery "group"—its physical case size and where the terminals connecting it to the car are located. The second is the battery capacity, expressed as "cold ranking amps" (CCA). This measurement indicates just how much electrical power the battery can produce. When comparing this rating on the new battery to your old one, read carefully! Some batteries show the CCA rating as measured at zero degrees Fahrenheit, while others show CCA at thirty-two degrees. The thirty-two degree rating will be a higher number, so be sure you are comparing "apples to apples."

A battery's capacity is rated by uniform testing procedures, but there can be differences in how the components inside are assembled. Heavy-duty, high-performance, and racing batteries are very strongly built, more resistant to temperature extremes and vibration from rough roads, and, accordingly, more expensive. Don't buy such a battery unless you really need it. Nor is it necessary to buy a 700 CCA battery if your car only needs 550 CCA. The cheapest may be the best, because that higher capacity battery has the same life expectancy as the cheaper one. It is true that having more electrical power available may allow your car to continue starting with a battery that is weakening, but once the battery begins to fail, it usually doesn't take long before it fails completely.

Time to Re-Tire!

"Your brakes stop the wheels, but the tires stop the car!"

— The Tire Industry Safety Council

We're serious about this. Not paying attention to tires is arguably the most common form of neglect among car owners. In over 300,000 cars tested nationwide by AAA, one out of three was found to have tires with incorrect air pressure, worn out tread, or other defects. Pretty scary when you realize—as you barrel along at eighty miles per hour in the rain, take curves and corners at speeds you probably shouldn't, and slam on the brakes as the rear end of a stopped vehicle looms up in front of you—that the only part of your three-thousand-pound car that actually touches the road is a portion of each tire about the size of a footprint! Like the failure of brakes or steering gear, failure of a tire while the car is moving can result in catastrophe.

Unlike batteries and oil, tires have no predictable number of miles that they can be expected to last. They are rated for treadwear, traction, and temperature, but treadwear rating is really just an indicator to help you compare one tire against another. The most important factors in determining your tires' useful life are in your hands: how well your tires and wheel alignment are maintained, and your driving habits.

The ratings for your tire are included in the information found on the sidewall. One set of characters tell the tire's size, shape, and construction. You will see a combina-tion of letters and numbers, such as "P225/70 R 15." The *P* indicates the load rating of the tire; *P* is for passenger car, *C* is for commercial, *T* is for temporary use only, *LT* is for light truck. "225" means that this tire is 225 millimeters wide when mounted and inflated. "70" indicates the profile of the tire. This tire's height is 70 percent of its

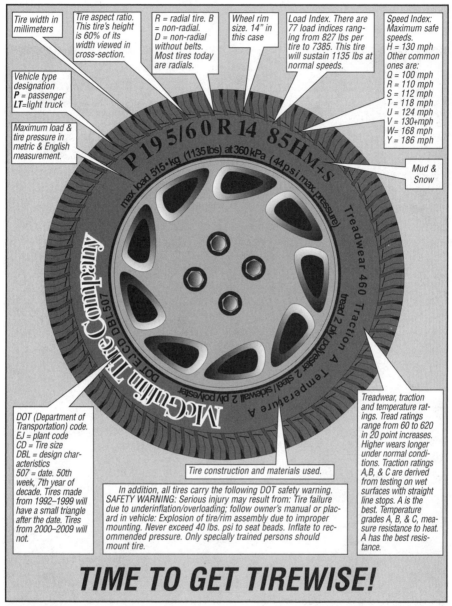

Tire width in millimeters

Tire aspect ratio. This tire's height is 60% of its width viewed in cross-section.

R = radial tire. B = non-radial. D = non-radial without belts. Most tires today are radials.

Wheel rim size. 14" in this case

Load Index. There are 77 load indices ranging from 827 lbs per tire to 7385. This tire will sustain 1135 lbs at normal speeds.

Speed Index: Maximum safe speeds. H = 130 mph Other common ones are: Q = 100 mph R = 110 mph S = 112 mph T = 118 mph U = 124 mph V = 130+mph W= 168 mph Y = 186 mph

Vehicle type designation P = passenger LT=light truck

Maximum load & tire pressure in metric & English measurement.

Mud & Snow

P 195/60 R 14 85HM+S

max load 515•kg (1135 lbs) at 360 kPa (44psi max pressure)

Treadwear 460 Traction A

tread 2 ply polyester 2 Steel / sidewall 2 ply polyester

Temperature A

McGriffin Tire Company

DOT EJ CD DBL 507

DOT (Department of Transportation) code. EJ = plant code CD = Tire size DBL = design characteristics 507 = date. 50th week, 7th year of decade. Tires made from 1992–1999 will have a small triangle after the date. Tires from 2000–2009 will not.

Tire construction and materials used.

In addition, all tires carry the following DOT safety warning. SAFETY WARNING: Serious injury may result from: Tire failure due to underinflation/overloading; follow owner's manual or placard in vehicle: Explosion of tire/rim assembly due to improper mounting. Never exceed 40 lbs. psi to seat beads. Inflate to recommended pressure. Only specially trained persons should mount tire.

Treadwear, traction and temperature ratings. Tread ratings range from 60 to 620 in 20 point increases. Higher wears longer under normal conditions. Traction ratings A,B, & C are derived from testing on wet surfaces with straight line stops. A is the best. Temperature grades A, B, & C, measure resistance to heat. A has the best resistance.

TIME TO GET TIREWISE!

Figure 3.6—Time to Get Tirewise!

width. A tire with a profile of 50 would be twice as wide as it is high. The *R* refers to the tire's construction: *R* for a radial tire, *B* for a non-radial belted tire, and *D* for a non-radial tire without belts in the tread area. The

WHAT SIZE TIRES SHOULD YOU BUY?

It is not uncommon to see cars and trucks with tires of a size different than that installed at the factory. This could happen for several reasons. Sometimes a tire that is narrower or has a higher profile is less expensive. On the other hand, wider tires and/or custom wheels may have been installed strictly for looks, or because the owner thought they would get "a better ride" with wider tires. In the case of a truck, very large diameter tires are sometimes installed to gain additional ground clearance and better traction for off-road driving.

While moving up or down by a single size graduation in the width or profile will generally do no harm, changing your tire size by any more than that may cause problems. Installing wider/larger tires decreases the clearance around them under the car, so the tires may rub against the wheel wells, steering, and suspension components, or even brake hoses!

Larger tires also weigh more, adding "unsprung weight" to the car's load; this will actually stiffen the ride. Unspring weight is the total of any components not supported by the vehicle's springs. Anything located between the spring and the road is unsprung. The more unspring weight, the harsher the ride. Suspension components in this area of the car are made to add as little unsprung weight as possible; they are sometimes even made from aluminum alloys. Watch the driver in one of those trucks with very large tires as it goes down the road, and you will see that it is anything but a smooth ride.

Custom wheels do not add or take away much unsprung weight, but they can cause problems. If the wheels are wider, then wider tires must be installed. To keep from rubbing brake hoses or suspension components in tight turning maneuvers, many people choose "deep-dish" reversed wheels that move the tires outward. Unfortunately, this solution has an unpleasant side effect. Moving the tires outward puts more load on the suspension and steering systems, shortening the life of wheel bearings and joints.

number "15" means that this tire is intended to be mounted on a 15-inch diameter wheel.

The traction and temperature ratings are listed on the sidewall as letters *A, B,* or *C. A* indicates the highest, or best, rating, *C* the lowest rating. Most radial tires carry a traction rating of *A* or *B,* with a temperature rating of *B* or *C.* It is unusual, and expensive, to find a tire having a temperature rating of *A.* Unless you are driving in extreme heat, you do not need to shop for these. If the tire is approved for good traction in mud and snow, the letters *M+S* will also appear on the sidewall. Most tires advertised as "all-season" will have this rating, along with a traction rating of *A.*

Figure 3.7—Not much to look at, but still hard at work, this 1990 Dodge Dakota has has only three owners and has logged over 133,000 miles. The summer sun has taken a toll on the paint job, but the truck is very sound without a speck of rust underneath! The current owner has just replaced the headliner and plans to have it repainted soon. Total investment for these repairs: less than two monthly car payments!

Figure 3.8—This '87 Pontiac Grand Am has accumulated over 180,000 trouble-free miles. The owner loves this little car and is planning to have it repainted soon. No plans to trade or sell it!

The treadwear rating listed on the sidewall is a point system. Typically, you will find numbers from around 150 for less expensive tires, to 360 or more for premium tires. While not meant to translate to a specific number of miles, each 10 points is equal to about three thousand miles of tread life in carefully controlled testing. It is a good indicator of how quickly one tire will wear out compared to another of the same size and construction. The difference is found mainly in the formula of rubber compounds with which the tires are made.

The tires that came on your car from the factory should have been carefully matched and evaluated to complement your car's handling and ride qualities. In fact, they will have a number with the letters "TPC," tire performance criteria, on the sidewall. This is the specific testing and standards that these tires meet. Unfortunately, only original equipment tires come with this TPC number. The only sure way to get an equivalent tire is to replace the originals with the exact same brand, size, and type. While you certainly do not have to do this, you should be aware that tires are engineered products, like many other components on your car. There is a difference in the ride quality and handling characteristics from one tire to another. While you would generally expect a more expensive tire to be "better," you might be disappointed to find that the more expensive tire is designed for sports-car type handling and has a very harsh, stiff, ride quality. Your local tire dealer can provide informed advice on replacing your tires. Take the time to discuss options and the qualities you are looking for. Get them to show you several tires, and compare their treadwear, traction, and temperature ratings against the price.

Nearly all cars have a tire placard affixed to the rear edge of the driver's door, the body pillar at the rear of the driver's door, or the inside of the glove box door. On the placard, you will find the size and type of tire with which the car was equipped from the factory. You will also find recommended tire pressures for both normal loads and maximum load conditions. Keeping your tires inflated to the recommended pressure will help you get the best combination of smooth ride and long tread life. Note that the recommended pressure on the tire placard is different from the maximum pressure found on the tire sidewall. The maximum pressure should be used only when carrying heavy loads. Never exceed it.

Checking your tire pressure should be added to your list of things to do weekly. We suggest this because it is normal for a tire to lose about a pound of pressure per month.

SPEEDOMETERS AND CRUISE CONTROL OPTIONS

Most modern speedometers are quite accurate. They are now electronic—no more speedometer cables. Today's odometer also is electronically operated. Yes, tire size will affect accuracy. Smaller diameter tires will cause both speed and mileage to read higher. Larger diameter tires will cause low readings, though usually not enough to get you a speeding ticket (unless we are talking thirty-inch off-road tires).

Did you know that, on most late models, your cruise control's set speed can be adjusted up or down one mile per hour at a time, just by quickly tapping the accelerate button (to speed up), or the coast button (to slow down)? It's a very convenient feature to use on the Interstate as traffic conditions change, if your car has it!

To find out about this, and to discover a lot of other useful (if somewhat obscure) information about the finer points of your car's features, do some good, enjoyable bedtime reading in your vehicle owner's manual!

Also, pressure in the tires fluctuates as the outside air temperature goes up or down. It takes only a few seconds to walk around the car and check each tire with an inexpensive tire gauge (available at any auto parts store). Neglecting tire pressure can result in shorter tire life, poor handling, and, at worst, a flat. When you have a flat may also be the only time you think about whether or not the spare tire has any air in it. So don't forget to check the spare! Sometimes this isn't easy, and you may have to take the spare tire out of its mount in order to reach the valve stem. (Davidson theorizes that this is how car makers can tell how many people actually check their spare. He says they count the letters complaining about the tire location!)

Wheel and tire troubles generally come in three forms: shimmy or vibration, premature wear, and blowouts. If you become aware of either of the first two and have the problem corrected, blowouts are not likely to happen—unless you are prone to running over large, sharp objects or railroad spikes!

If you feel a shimmy or "waddling" sensation at low speeds, just before stopping or when you first begin to move after a stop, it usually means that one of your tires has a separated belt, a condition that will cause the tire to fail. Have your tires carefully inspected as soon as possible. It should be noted that a bit of this sort of waddle is normal when first starting out in the morning, especially in cold weather.

A shimmy or vibration at highway speeds, especially one that phases in and out as you speed up or slow down, is likely to be caused by imbalance, but it could also signal that a tire is beginning to come apart. Furthermore, you can tell where the vibration is located if you tune in carefully. A shimmy or vibration that is felt directly in the steering wheel is almost always from a front wheel, while one that is felt in the floor or seat, but not in the steering wheel, is most likely originating from a rear wheel. Having the tires balanced will correct the condition if you don't have a tire problem. If you do, the technician should be able to see a tire's impending failure as the wheel spins during balancing.

Wheel balancing is also effective in preventing one type of premature tire wear. A tire that is run for long periods in an imbalanced condition will develop a scalloped wear pattern in parts of the tire, commonly called "cupping." Incorrect wheel alignment (that is, the angle at which the tires contact the road), can also cause premature wear. It usually causes one side to wear more quickly than the rest of the tire. Wheels can come out of alignment during the normal course of driving, especially over rough surfaces and potholes, but misalignment can also result from loose, worn, or bent steering and suspension components and joints. Note, however, that wheel alignment will not correct a vibration or shimmy!

Rotate your tires as recommended in your maintenance schedule (usually every 7,500 miles for front-wheel-drive cars), and keep an eye on the depth of the tread remaining on them. You can check the tread depth yourself with an inexpensive depth gauge (available at auto parts stores), or just stick the tip of a small ruler in the tread grooves. Measure the depth at the second groove from the outer edge of the tread. Most new tires have tread depth of $10/32$ to $12/32$ of an inch. When the tread wears to about $4/32$, you should begin thinking that you may not have to rotate these particular tires again. You may also notice that you have to drive more slowly on wet roads to maintain good steering control.

When the tread depth reaches $3/32$, you need to start checking the sports section of the Sunday paper to see who has tires on sale. (They always run the tire and auto parts ads in the sports section; it's a guy thing, we suppose!) At $2/32$, the tread wear indicators built into the tire will be even with the top of the tread, giving the appearance of a bald stripe every eight inches or so around the tire. Continuing to run the tire in this state is not only dangerous, it can get you a traffic ticket most anywhere in the United States. If you should have an accident, you may also be found to have been negligent, which could result in your being found at fault, even if you really were not! So you see, there are lots of good reasons to re-tire when it's time.

Brakes and Wheel Bearings

Brakes are critically important because, along with the tires, they are the most important safety equipment on your car. Wheel bearings provide the means for your wheels to turn mile after mile with minimal friction and maximum economy.

Years ago, brake systems were less complicated, but the basic brake system components haven't changed much in the last thirty years, except for the material from which the brake linings are made. Brake pad and shoe lining material no longer contains asbestos, which used to be the main ingredient. The biggest difference in design came about with the introduction of antilock brake systems (ABS). ABS first appeared on a few American luxury cars in the late 1960s. A somewhat crude system, these early versions provided some antilock control on the rear wheels only. Today, most new vehicles have computer-controlled ABS, and many have antilock control on all four wheels. A few even have the ABS combined with a traction control system. The traction control system takes control of the accelerator in emergency maneuvers, to help keep the car from going into a skid.

All your brakes ask of you is that you replace worn components and not neglect brake fluid changes. In return, they will not let you down when you call upon them in an emergency stop on a hot afternoon in rush hour! Inspecting the brake system periodically is important to be sure that you will have all the braking efficiency with which your car was designed when you need it most. Ask for a brake inspection at the same time you have your tires rotated, usually every six months or 7,500 miles, whichever comes first. Many shops will provide this service free of charge.

Most late model cars and trucks are equipped with wear indicators built into the disc pads on the front brakes. The rear brakes generally do not have wear indicators, because the front brakes normally wear out more quickly, sometimes four times as quickly on a front-wheel-drive car. These indicators are small strips of metal that are designed to touch the rotor when the pads are nearly worn out. When they do, a terrible, "nails on the chalkboard" sort of squeal is heard. The wear indicators generally squeal when the brakes are not applied, so they are quite annoying—and that is exactly their job! They are designed to make you inspect your brakes while there is still enough lining left on the pads to prevent major damage or loss of safety.

Along with the brakes, inspect those wheel bearings that silently carry the car's weight and allow the wheels to turn freely. Years ago, all wheel bearings needed to be periodically removed, cleaned, and repacked with grease in order to keep doing their work efficiently. This was generally done when the brakes were relined. Beginning around 1980, sealed ball-type bearings started to become commonplace. These sealed bearings are now used in the vast majority of cars, and in a few light trucks, as well. Sealed bearings require no periodic maintenance, except to check them for looseness. When they become loose and/or noisy, the bearing assembly must be replaced.

Protecting Your Automotive Investment

To recap, you should make a point to read your owner's manual and maintenance schedules carefully; make sure that you are using the correct fuel and the right fluids for all of your car's systems; and make it your business to understand the way your car works and tend to its needs on a regular basis. If you need more help to learn about your car, or cars in general, than you find in this book, see if your local vocational school or community college offers a program such as "consumer auto repair." As with a person you care about, understanding your car makes it pleasant to spend time together! It can actually be fun to go outside on a brisk, beautiful morning and spend a few minutes puttering under the hood and around the car. At least, we think so.

As you drive, turn off that exotic sound system for a minute and listen. Listen to everything that makes a sound. What does it sound like as you crank the engine in the morning? What kind of noises do you hear when you negotiate turns? What does the steering feel like in tight maneuvers? What happens, and what do you hear, as you encounter different kinds of bumps in the road? Arm yourself with the knowledge of what your car sounds and feels like when everything is okay, so you can describe what has changed if something doesn't seem right. The days of the "tune-up" as a cure-all for sluggishness and general performance maladies are gone. Today, your repair technician must perform pinpoint diagnostic tests on what have become extremely sophisticated and complex systems to determine the nature of many problems. In nearly every case, though, the technician must rely upon you to point him or her in the right direction.

If something goes wrong while your car is still covered under warranty, you may need to produce proof that your maintenance chores and inspections were performed on time. For the most part, it is acceptable for you to perform the maintenance yourself, or have it done at any repair facility you choose. Questions arise when a failure occurs that could have been the result of neglected fluid changes or other services required at specific times or odometer readings. For example, you would not need to produce proof of oil changes if your engine's head gasket began leaking. But if the oil could not circulate through the engine because of excessive sludge and the bearings burned up, proof of oil changes might well be necessary to protect your warranty. All you need to do is keep consistent records of your service as it is done. If you have your routine maintenance done for you, you should keep the work orders. If you change your own fluids, you should keep track of your service with a written log book and also save the receipts when you buy fluids, filters, etc.

Getting off to a good start with this new car of yours can ensure many, many miles of enjoyment. We want to help you enjoy every mile of at least the 300,000 you will spend with this machine you love. In the next chapter, we will talk some more about things that can go wrong and how to communicate with your technician or service advisor to make your repair experiences as painless as possible.

CHAPTER 4

GETTING TO KNOW YOUR CAR

The Normal Car: What's Normal, Anyway?

Now that we've checked the oil and other fluids, and examined and, perhaps, replaced the battery and tires, it's time to take a good look at your new-to-you car from the point of view of driving it every day. Not only do you have to depend on this car, but you have to live with it, too. Adjusting yourself to its individual quirks and oddities, and not getting panicked about them, is an important first step in having a good relationship with your new car.

A particular Lexus model released some years ago would make a noise that sounded like a load of loose lumber in the trunk whenever it went over a bump. There was no fix for this condition. Every Lexus of this particular model would do it. Some owners found this particular quirk extremely annoying—to the point where they requested the manufacturer to buy the vehicle back— while many others were not bothered by it at all. Indeed, some owners apparently never even noticed it.

While each model of car and truck has characteristics different from others from the same manufacturer, it is also true that there

Once used for transporting guests to the airport, this '91 Chevrolet Astro is lovingly maintained by its second owner and looks showroom fresh inside and out at 154,000 miles. He takes pride in this van, as well as his award-winning 68 Mustang!

can be differences between two identical models. One car's seat may feel a bit different, or one may idle more smoothly. One might generate a long repair history, while the one that came off the line right after it will run happily for 300,000 miles or more with few difficulties.

It is important to realize that, just as with people, there are certain quirks found in each car that can be accepted and considered normal. Maybe computers and robotics in car production will change this some-day, but somehow, we think not. Obviously, some quirks and conditions are more annoying than others, both in people and in cars, with some conditions outside of the realm of acceptability.

The time to notice those unacceptable quirks, of course, is before you buy the car. In chapter two we give you a complete rundown on how to avoid unacceptable cars with a checklist of questions to ask, and things to do and listen for when test-driving. Again, do not accept a car or truck that was in any way unsatisfactory during the test-drive. If the salesperson assures you that a problem you pointed out can be correct-ed, be skeptical. Be sure to have them document the problem with a "due slip," and get a copy of it. That way, you have evidence that they promised correction of this condition as part of the sales agreement. If the problem is not corrected, cancel the sale. Simple as that. There are plenty of other cars out there.

After taking possession of your newly purchased car or truck—during the honeymoon period—drive it a lot, under different conditions, and pay careful attention to its characteristics. Turn off the sound system. Drive it with the windows up and with the windows down. Listen carefully. Feel the car through the steering wheel, accelerator, shift lever, and the seat of your pants. Try out every feature. Knowing how your vehicle normally feels, rides, and performs will provide a basis for comparison and will help you sense small changes. This is important, because problems usually begin to show themselves in the form of small variations from "normal." A low tire will cause a swaying sensation in curves and corners before it goes completely flat. Most air conditioners begin cooling less efficiently before they quit altogether. Of course, there are exceptions, some of which we have mentioned, but many impending problems can be detected early with careful attention to what is normal for this vehicle.

Once you have driven your car for awhile, you will become aware of its personality traits. Perhaps there is a wind noise that only happens under certain conditions, or a little creak or squeak in the dash every morning until the car is warmed up. These could be examples of quirks that are not correctable. Generally, a small noise, vibration, or change in feel after a few hundred miles, or one that is first apparent at a seasonal change such as the onset of cold weather, is nothing to get excited about.

One of Roy's friends has a car that has always made a hissing noise from under the center of the dash, but only on hard braking from speeds of over thirty miles per hour. In over 133,000 miles and nearly nine years, no one—including, we have to say, Roy—has ever found the cause and no functional problem has ever occurred. It is an interesting little quirk, to say the least!

On the other hand, an engine that stalls when cold and a transmission that refuses to shift smoothly are genuine problems that should not be tolerated. If a visit to the repair shop results in a response such as "no problem found" or "they all do it," get a second opinion. The next step is to try to get your hands on an identical model, even if you have to rent one for the day. If this noise (or whatever) is truly characteristic of this model, your dealer's service manager or sales manager should be willing to arrange for you to drive a similar car or truck. If they all really do it, the comparison car will do it, too!

Back in the old days, the Packard Motor Car Company always included in their ads the phrase, "Ask the Man Who Owns One." Of course, this (wholly politically incorrect) statement was intended to make the reader believe that anyone who owned a Packard would give it a glowing endorsement. That's still pretty good advice, even though the last Packard rolled off the line in 1955. Be careful, though: not only do car owners of all genders tend to be subjective about their cars, but also two cars that look identical may be quite different underneath the hood. Systems and components are often redesigned, sometimes within the same model year. This can result in different sounds, performance, feel, and fuel economy in the same year and model.

A funny thing, though. The appearance of minor noises and other conditions often brings out a quirk of human nature. Once we become aware of a minor annoyance, such as a bump that happens at one exact throttle position, we have a tendency to reproduce that exact throttle position repeatedly, rather than simply pushing through the spot where the bump occurs. During Roy's days in dealership service departments, there was a particular model of diesel-powered Cadillac with just such a normal trait. If you held the accelerator just so while climbing a hill, the transmission would hunt back and forth between second and third gear. There was no fix for this condition and it offered no safety compromise; it was normal for that model. Roy says:

> All we could do was adjust the throttle position where it happened. But the transmission would always hunt at some combination of speed and load, no matter what. I was amazed by the number of drivers who would always manage to find that exact accelerator position within a few days, even after we had changed it!

If your new car should surprise you with a distasteful little trait of some sort that really is normal for that model, you have a few choices. If the quirk is something you can live with on an otherwise perfect car, peaceful co-existence is the easiest and most satisfactory answer. Push through that spot on the accelerator that makes the transmission hunt! Cover up that little creak in the dash on a cold morning by turning on that beautiful high-tech stereo and listening to your favorite music. (At reasonable volume, of course. If that doesn't overcome the noise, it is *not* normal!) As a second choice, you can always try to find someone who can overcome the annoying condition by making some modifica-

BEWARE OF FUEL ECONOMY QUOTES!

Unexpectedly poor fuel economy is a frequently noted "quirk." But it is perfectly natural for two identical models to get very different fuel economy due to all of those factors we've discussed, including how, where, and by whom the car is driven. Those EPA fuel economy ratings on your car's window sticker might not be accurate for your driving conditions.

Also, be aware that Internet-quoted fuel economy numbers may be way off the mark! In their zeal to provide the maximum level of comparison information, dot-com car services sometimes compute their own fuel economy estimates by averaging the EPA fuel economy ratings of several "similar" vehicles. So, especially with earlier model years, vehicles that were not even tested by the EPA may be supplied with economy ratings! Sadly, even some dealerships rely on these figures and distribute them to customers in their information sheets for a particular make and model. If you are looking at a used car, you can check the EPA's Web site yourself, or contact a representative of the auto buyer's guide you are using and ask how the fuel economy figures for that model were determined. Be skeptical, and don't expect miracles.

If your engine is truly using too much fuel, there is certain to be a change in its drivability, such as sluggish performance, excessive spark knock, increased operating temperature, or higher than normal engine speeds.

tion or adjustment within the limits of the manufacturer's warranty. If certain major changes are made, the warranty for some components, or even the whole car, may be voided. That is the last thing you want to happen.

Finally, you have the option to seek assistance under the lemon laws, but that should truly be a last resort, and is only an option for a substantial, serious defect. We will discuss the lemon laws in a later chapter. Like we said above, get a second opinion and drive a comparison vehicle before you decide on any course of action.

About Today's Cars

Today's car is nothing like the car of ten years ago. It may have a tire in each corner, an engine, and a nice place to sit, but under the hood, and nearly everywhere else, it is a different animal. As computers, digitalization, and the Internet have transformed the world we live in, so has the digital age transformed the car. It's not your father's Oldsmobile, Chrysler, Crown Vic, or any other "dumb" car your father or mother might have had. Today's cars are "smart." They have to be—it's the law!

During the last decade, car designers and engineers were confronted by tough challenges. New laws insisted on higher standards for safety, cleaner-running engines, and increased fuel economy—all at same time. So car makers had to produce cars that could meet all these criteria and still meet the needs and desires of the public.

The old way of designing cars often resulted in trading a gain in one area, such as reduced exhaust emissions, for a loss in another, such as performance. Car manufacturers scrapped those plans and began designing systems that could work together in harmony and without loss of features. Fortunately, the digital technology they needed was there to help them.

The period between 1979 and 1991 saw the gradual end of heavy engines, the carburetor, and "bolt-on" emission controls. It saw the emergence of the compact, lightweight, fuel-efficient machines of today. The rest of the decade saw design improvements and the introduction of new materials and components to accomplish improved performance and durability. More and more functions are controlled by computer, and the end is not in sight. Now, engines run smoother, valve trains last longer, fuel economy is better, and accessories and features continue to grow.

Smart cars, indeed! If a mechanic of yesteryear were to be suddenly transported from the 1960s into the new century there would be little that he would recognize beyond the basic mechanical engine components, the transmission (still recognizable, but now shifted electronically), and the non-electronic brake parts and running gear. He'd probably have a funny haircut and nice long sideburns, too.

Roy laments,

> *It would be like putting Ben Franklin in modern Philadelphia. He might recognize Independence Hall if he stumbled across it, but he wouldn't have the slightest idea where to buy ink for his quill pen, or where to get a sherry flip (which he would probably need). That is, if he didn't have a heart attack or get run over by a bus. An old-time mechanic would be just as lost under the hood of a modern car. He knows the basic engine is still in there somewhere, but everything that controls it is different!*

This rest of this chapter and the next will describe for you both the car's mechanical systems and the electronic systems that govern them. But first things first.

Learning the Lingo

It is a good idea to understand some of the more common nomenclature used to describe today's complex automotive systems and the problems that sometimes arise. It will help later on when you talk to service technicians, and even other drivers, about your car. There is a universal language of "car talk" (with appropriate apologies to the Magliozzi brothers), and you will be amazed at how your reputation as a car maven will soar when you use it.

STOP

INCORRECT TIRE PRESSURE

Incorrect tire pressure can cause or aggravate a number of conditions; harshness, rattles, leading and torque steer. To make sure your car rides and handles as it should, use only the tire sizes and pressures recommended on the car's tire placard. The placard is usually found on the rear edge of the driver's door or the adjacent body pillar, but it may also be located on the inside of the glove box door.

There's another reason to use it, too. Unfortunately, the very descriptive words often used to try to communicate a problem clearly may mean different things to different people. One person may say, "the engine won't turn over" and mean that the starter motor doesn't run when the key is turned. Another person may say the exact same thing and mean that the starter is operating, but the engine won't start! So it really helps to understand some basic terms. Let's review the more common standard terms used to describe what is going on (or not) under the hood and elsewhere.

The Normal Car's Basic Functions

- *Crank:* What the starter motor does when you turn the key. When the starter is operating normally, the engine is *cranking* as it should, whether or not it starts successfully.

- *Start:* When the engine begins running on its own without help from the starter motor, it has *started.* If it then stops running, this is called *stalling.*

- *Charging:* When the vehicle's battery is being replenished as the engine runs, it is *charging.* The system that accomplishes this function includes the alternator and is called the charging system. If the battery is weak, lacking enough power to crank the engine and operate accessories, it is referred to as *low in charge,* or *discharged.*

Terms Describing Problems

- *Understeer:* When a vehicle is steered into a turn, it has *understeer* if it tends to go straight, rather than follow the steering command. In auto racing, this is also referred to as *pushing,* or simply *tight.*

- *Oversteer:* As a vehicle enters a turn, if the rear slides outward and causes it to skid and spin around, the vehicle is *oversteering.* In racing, this is called *loose handling.*

- *Leading:* A vehicle that steers itself to one side is said to *lead.* A slight lead to one side under certain conditions may be caused by the road you are driving on, or by an unusual load in the vehicle. Leading is most often caused by uneven tire pressures, the need for wheel alignment, a defective tire, or by worn or damaged suspension parts.

- *Pulling:* A drift or steer to one side when braking is called a *pull.* Remember the rule of thumb: steering leads, brakes pull. Pulling is most often associated with problems in the brake system (not the ABS), but may also be caused by uneven tire pressures, uneven tire sizes, or uneven tire wear. Have these items checked before brake diagnosis is undertaken.

- *Torque steer:* A drift or lead to one side when accelerating or slowing down is called *torque steer.* It is more common on front-wheel drive vehicles and rear-wheel drives with limited slip differentials, and some models are more susceptible to it than others (see chapter five). Torque steer can come from worn suspension parts, but is usually caused by a difference between the left and right tires on the driving axle. Uneven tire pressures, different tire sizes, or differences in the wear on tires can all cause this condition. Slight torque steer is unavoidable on certain models, but an excessive amount can be dangerous and must not be tolerated.

- *Vibration:* This is a rhythmic, repetitive shaking or buzzing sensation. A low-frequency vibration may be relatively slow and seem to be related to the speed of the wheels and tires; a rapid, high-frequency vibration is more of a buzz at engine speed and may even cause a sensation of pressure inside the passenger compartment. Local vibrations in the dashboard area can usually be easily located. Place a finger between the suspected vibration location and the dashboard component next to it. If either of the two adjacent pieces is vibrating, you will immediately feel it.

- *Harshness:* This refers to the sensation of a stiff, harder than normal ride quality, especially on bumpy surfaces. Oversized or overinflated tires are the usual suspects, but frozen or stiff shock absorbers may also cause harshness.

- *Numerous, annoying noises:* Noise descriptions run the gamut. Despite the best attempts to standardize noise descriptions, a "squeak" to one person is a "chirp" to another, and one's "thump" may be another's "clunk"! The best advice with noises is to take a test drive and actually point out the noise to the technician, service writer, or other professional.

Terms for Engine Driveability (Performance) Problems

- *Missing* or *misfire:* A combined vibration and associated lack of power caused by one or more cylinders not contributing its fair share of power to the engine is referred to as *missing* or *misfire.* A constant, steady misfire is referred to as a *dead miss,* while an occasional misfire that causes a momentary jump or bump is called an *intermittent miss* or *dropping a cylinder.* Although misfire is usually associated with the engine's ignition system, other causes are fuel delivery and basic mechanical problems. Most cars and trucks since 1996 are equipped with a self-diagnostic system called On-Board Diagnostics, second generation (OBD II), which we will discuss later. That system can detect misfires and even tell which cylinder(s) they are coming from. (By the way, when someone tells you, "Hey, your engine is missing," do not say, "No, it's not, it's right under the hood. I saw it there when I checked the oil!" We've all heard that one.)

- *Surging* and *chuggle:* While these terms don't sound very scientific, they are commonly used in auto makers' technical service bulletins to describe a jerking motion, as though the car were running over a railroad crossing. *Surging* or *chuggle* is caused by uneven power delivery from the engine—usually resulting from uneven fuel delivery (itself the result of dirty fuel injectors or carbon build-up), or incorrect calibration of emission controls or sensors.

- *Hesitation* or *stumble:* This is an apt description of what happens when you try to accelerate, but there is a momentary time lag before the engine begins to pick up speed. It is usually most noticeable when you take off from a traffic light or stop sign; the engine may pop, or jump a bit, then finally speed up. Hesitation is usually caused by a faulty sensor in the fuel delivery system or by defective ignition components. Stumble during warm-up, after the engine is first started, may also be caused by carbon build-up.

- *Lag* or *bog:* This unpleasant condition, also referred to as a *flat spot,* is a sensation of no response to increased throttle. You push the accelerator, but the car just seems to sit there. Sometimes you can push through a flat spot and reach a point when the engine will pick up power suddenly. The causes are generally the same as those listed for hesitation or stumble, and for surging or chuggle.

When you take your car in for service the most important single thing that you can do is to clearly communicate your car's problems and help lead the service technician down the path to correct diagnosis and resolution. Learning to "talk the talk"—at least a little—is valuable for making your job, and theirs, easier. It is well worth the effort!

The Normal Car's Basic Systems

The Engine

Your car's engine is actually a large air pump with several support systems added to it. If you were to take away the support systems, and turn the engine over using an external power source, that is just what you would have: an air pump. If the engine were not designed to be an efficient air pump, it would not breathe properly as it ran. The power it could deliver would be greatly reduced and it would run hotter. In addition, it would waste fuel and be what the EPA defines as a "gross air polluter." Poorly maintained engines and those in need of repair will lose their ability to function efficiently, sometimes in an alarmingly short number of miles. The signs of trouble are substantially reduced fuel economy, sluggish performance, higher than normal operating temperature, and excessive oil consumption.

Some years ago, we would have advised that blue or black smoke might appear at the tailpipe to show engine problems: blue would indicate burning motor oil; heavy whitish-blue would indicate burning automatic transmission fluid; and black would indicate flooding from too much fuel. Today, however, engines are equipped with efficient emission control devices that cover up a multitude of mechanical sins by burning off smoke before it even reaches the tailpipe. Certainly, a problem is indicated if smoke *is* seen, but lack of smoke does not necessarily mean that everything is okay.

Be aware that the presence of white vapor and water dripping at the tailpipe on cold mornings is normal, unless the engine's cooling system frequently requires topping off. In that case, it may indicate a leaking head gasket or other internal leak. A serious internal leak will usually result in white vapor that continues even after the engine is fully warmed up.

The engine and its supporting systems can be divided into seven categories, not counting the accessories mounted to the engine (such as the alternator, air conditioning compressor, and power steering pump). The categories are:

- Basic mechanical components: These are the "pumping" components of the engine, such as the pistons, crankshaft, flywheel, and harmonic balancer, and the "breathing" components, such as valves, camshaft, pushrods, and rocker arms. In addition, bearings are needed to support the rotating parts.

- Lubricating system: The components of this system include the oil pump, oil filter, oil passages, and, of course, the lifeblood—motor oil. The lubricating system is responsible for reducing friction and helps cool the internal parts.

- Cooling system: Enough heat to warm a two-bedroom home in winter must be removed from a modern V-8 engine! This system includes the water pump, radiator, cooling passages, hoses, and antifreeze/coolant.

- Fuel system: Components responsible for fuel delivery are the fuel pump, fuel filter, pressure regulator, fuel injector(s), rail, hoses, and the sensors and electronic controls that meter the exact amount of fuel needed for optimum economy and driveability. The air filter may also be considered a fuel system component as it must allow sufficient airflow to provide the correct air/fuel mixture.

- Ignition system: This team delivers a spark to ignite the compressed gasoline and air mixture at just the right time. Components include the spark plugs, spark plug wires (still present on most models), distributor (not present on many models), ignition coil, and electronic sensors and controls.

- Exhaust system: Components of this system provide the outlet path for the engine's burned air/fuel mixture. The system must also provide resistance to the flow of exhaust, and create a carefully engineered back pressure in order to make some engine controls, such as the exhaust gas re-circulating (EGR) valve, function properly. If the system is leaking or tampered with, engine performance will be impaired, not improved! Included on most models are the exhaust manifold, exhaust pipe(s), muffler(s), and tailpipe(s). No, we did not forget the catalytic converter. Read on!

- Emission control components: On the modern engine, it is difficult to tell just where emission controls leave off and fuel and ignition controls begin, because they are so interrelated. The days when emission controls were removable, bolted-on parts are long gone! The modern engine uses the same electronic sensors, controls, and engine computer (power train control module, or PCM) for the fuel, ignition, and emission control systems. In addition, there are mechanical components to provide "clean-up" of specific types of emissions, such as the positive crankcase ventilation (PCV) valve, the EGR valve, and, yes, the catalytic converter.

Now, let's look at how the engine functions. The fuel system supplies fuel, which is mixed with air and drawn into the engine's cylinders, one at a time, through the intake valves. Inside the cylinder, the piston is pushed upward by the crankshaft and connecting rod and the air/fuel mixture is compressed. Then, the ignition system provides a spark to ignite the mixture. As the mixture burns, it expands rapidly and pushes the piston downward, rotating the crankshaft. The burned air/fuel mixture is pumped out of the engine through the exhaust valves and into the exhaust manifold, then on to the catalytic converter, muffler, and tailpipe. This cycle of intake, compression, power, and exhaust is repeated over and over in every cylinder of the engine.

To make sure that the valves operate in sync with the movement of the pistons, a timing mechanism must be used. It may be driven by a reinforced rubber belt, a chain, or a series of gears. The belt mechanism is the only one that does not require lubrication. A rubber timing belt will generally require replacement at 60,000 or 90,000 miles, depending on its design. A chain will usually be loose and ready for retirement at about 100,000 miles. This item is probably *not* listed on your owner's maintenance schedule, but you will be surprised at how much pep your engine will recover when it is replaced! The gear-driven mechanism may well last the entire life of the engine. A variety of factors are considered by the engineers when deciding what type of timing mechanism to use, including noise, serviceability, exposure to heat, availability of lubrication, and of course, price. We'll bet you can guess which one is the cheapest and which is the most costly to build!

The Drive Train

The drive train consists of the transmission, differential, and drive axles (which carry the power to the vehicle's drive wheels). On cars and trucks with a manual transmission, a clutch will be found between the engine and transmission. On vehicles equipped with automatic transmissions, the functions of the clutch are built into the transmission components. It is the job of the clutch, or its equivalent, to provide smooth

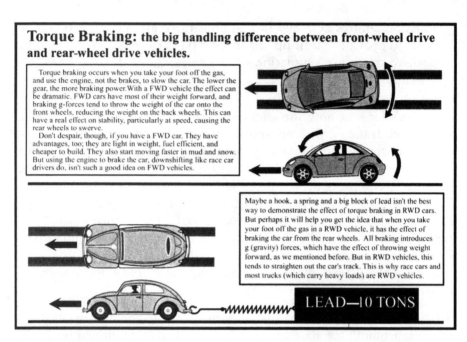

Torque Braking: the big handling difference between front-wheel drive and rear-wheel drive vehicles.

Torque braking occurs when you take your foot off the gas, and use the engine, not the brakes, to slow the car. The lower the gear, the more braking power. With a FWD vehicle the effect can be dramatic. FWD cars have most of their weight forward, and braking g-forces tend to throw the weight of the car onto the front wheels, reducing the weight on the back wheels. This can have a real effect on stability, particularly at speed, causing the rear wheels to swerve.

Don't despair, though, if you have a FWD car. They have advantages, too; they are light in weight, fuel efficient, and cheaper to build. They also start moving faster in mud and snow. But using the engine to brake the car, downshifting like race car drivers do, isn't such a good idea on FWD vehicles.

Maybe a hook, a spring and a big block of lead isn't the best way to demonstrate the effect of torque braking in RWD cars. But perhaps it will help you get the idea that when you take your foot off the gas in a RWD vehicle, it has the effect of braking the car from the rear wheels. All braking introduces g (gravity) forces, which have the effect of throwing weight forward, as we mentioned before. But in RWD vehicles, this tends to straighten out the car's track. This is why race cars and most trucks (which carry heavy loads) are RWD vehicles.

LEAD—10 TONS

Figure 4.1—Although front-wheel drive cars can move easily on ice and snow, they are often harder to stop safely. All of the engine's torque braking is applied to the front wheels as the driver tries to slow down, so the rear of the car tries to overrun the front and slips sideways. With rear wheel drive, the torque braking on the rear wheels tends to pull back as the car slows and helps to hold things straight.

engagement of the power from the engine to the transmission, and to disengage the power when changing gears.

The transmission carries power to the differential and provides the vehicle with the ability to travel at different speeds at a given engine speed. In the lower gears, the engine turns faster than the transmission's output speed to the differential. The engine's pulling power is increased, but top speed is decreased. In high gear, the transmission output speed is the same as the engine speed. In overdrive, the engine turns slower than the transmission output speed. This gear provides more fuel economy for cruising, but reduced pulling power. Overdrive is generally used only after reaching cruising speeds of over forty-five mph.

The differential provides further reduction in turning speed to improve pulling power, but it has an even more important job. The differential also divides the pulling power, called *torque,* between the left and right drive axles, which are attached to the wheels. In this way, the wheels are

Figure 4.2—The MacPherson strut suspension system is used almost universally on front-wheel drive vehicles.

much less likely to slip and cause tire wear as the vehicle turns, and safety is much improved.

There are three basic drive train configurations: front-wheel drive, rear-wheel drive, and four-wheel drive. Front-wheel drive is, by far, the most popular drive train used in automobiles today. There are three main reasons for this: front-wheel drive is very fuel efficient, light in weight, and cheaper to build. It also takes up the least amount of space, so it lends itself to the current designs, which use more space for the passenger compartment and less for the engine room.

With front-wheel drive, the transmission and differential are contained in a single housing assembly and are called a *transaxle.* The transaxle is mounted alongside the engine under the hood. The car is pulled along by the front wheels, which makes it easier to get moving in slippery conditions, such as mud and snow.

However, there is a negative side to this feature. When decelerating, all of the engine's braking power, called *torque braking,* is applied to the front wheels of a car that is already nose-heavy by design. The engine braking causes a pushing action at the front end, which causes the rear of the car to slip sideways and go into a skid quite easily. This tendency is made even worse if the front tires have better traction than the rear. That's right: believe it or not, it is safer to have your best tires on the rear of a front-wheel drive car! Also, never mount all-weather tires on the front wheels only.

Rear-wheel drive is still the most popular for two-wheel drive trucks and sport utility vehicles (SUVs). Racing cars use rear-wheel drive almost exclusively. It is a heavier system than front-wheel drive, because the transmission and differential are laid out differently. Additional components must also be used, including a driveshaft to connect the transmission (mounted behind the engine) to the differential (mounted between the rear wheels). In addition, the differential must be mounted in a separate housing.

However, this design moves the transmission and differential out of the engine compartment and provides engineers with more space under the hood. More and larger gears can be used—a perfect combination for heavy-duty use in vehicles where fuel efficiency is not quite so important. Naturally, the driving characteristics of a rear-wheel drive vehicle are opposite of front-wheel drive: it's harder to get going in the snow, but easier and safer to slow down on slippery surfaces. In fact, rear-wheel drive vehicles are generally easier to control all around. It is no accident that police departments continue to buy rear-wheel drive cars despite the fact that there are fewer choices every year.

Four-wheel drive is just that: there are two differentials, so power is distributed to both front and rear wheels. A transfer case is added behind the transmission to split the engine's power to two driveshafts. This type of drive train was originally designed for off-road use in military vehicles. Today, there are several varieties of four-wheel drive systems. They range from manually engaged units to electronically controlled systems that engage automatically, commonly called "all-wheel drive."

Since traction is improved over two-wheel drive, four-wheel drive would seem to be the best of all for safety, but that is not necessarily true. In fact, insurance company statistics indicate that four-wheel drive vehicles are as likely to go out of control as any other type of vehicle. Part of the problem may be that drivers of four-wheel drive vehicles can have a false sense of security and feel invincible.

There are a few drawbacks to four-wheel drive that may make you think twice about buying it if you really don't need to. There are more moving parts, as well as more weight. (Remember "unsprung weight" from chapter three?) So, vibration, noise, and harshness of ride will all tend to increase. Four-wheel drive vehicles also tend to wear tires more quickly than their two-wheel drive counterparts, and require more maintenance, especially if they are used off-road.

Suspension and Steering

Now, let's discuss the suspension and steering systems that, along with the tires, control the ride quality and handling in any car. Although there are lots of suspension system designs, only three basic types are commonly used on cars and light trucks.

The first is found almost entirely on rear-wheel drive platforms, and was introduced by Chrysler Corporation in 1949. It consists of a solid rear axle, supported either by leaf or coil springs, and independent front suspension, using coil springs and control arms (one upper and one lower

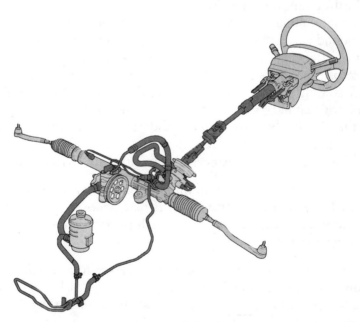

Figure 4.3—The rack-and-pinion steering design is widely used because of its light weight and few moving parts.

on each side). To provide good handling and steering control, the upper arm is shorter than the lower. For this reason, this type of suspension is often called SALA, for "short arm, long arm." To provide smoother ride quality, the control arms are made of lightweight stamped steel. Some more expensive imports use aluminum alloy control arms. In this way, unsprung weight is held to a minimum, allowing the arms to move easily over bumps without transferring the bump motion through the rest of the car. The solid rear axle is heavy, however, and one side cannot move independently of the other. This results in a harsher ride. That is one reason this design is most commonly used on trucks and SUVs today, and one more reason that front-wheel drive has become so popular: the solid, heavy rear axle is not needed.

The second design is used extensively on four-wheel drive vehicles. Although independent suspension is used on a few models, most four-wheel drives use solid axles front and rear, making for heavy-duty, durable components—and a ride to match! Lots of unsprung weight with this one. These systems may be supported by leaf springs or coil springs with trailing arms to control front-to-rear axle movement.

The third design, the MacPherson strut, is by far the most commonly used for late model cars, and in a few light trucks as well. Struts have been around since 1953, but they did not find wide use in automobiles until fuel efficiency, and hence size and weight, became an issue in the mid-1970s. The struts are similar in shape and function to those used on aircraft landing gear, and are used to replace the functions of the upper control arms and carry most of the load. The strut package, which contains the shock absorber, spring, and the spindle that the wheel is mounted on, is light, compact, and very strong. Only a small, lightweight lower control arm is needed to control side-to-side movement.

Struts are almost always used in the front on front-wheel drive vehicles, and may be used in the rear as well. If rear struts are not used, a light, semi-independent rear axle is often used. It is supported using coil springs and trailing arms.

Automotive engineers work constantly to make steering safer and more efficient. The "conventional" steering gear, using a gearbox mounted on the frame rail and a series of linkage rods to connect the wheels to the gearbox, is nearly obsolete. It is used today on only a few truck and SUV models. Today's steering gear of choice is the rack-and-pinion system, which grows more popular each year.

Originally intended as a means to reduce weight and take up less space in the engine room on fuel-efficient compact cars, the rack and pinion is now used on most cars, as well as many light truck and SUV models. The entire system consists of only three components: the rack-and-pinion gearbox, and two tie rods that connect it to the wheels. In addition to the benefits of its compact size and light weight, there are fewer joints to wear out.

Steering and suspension system components are carefully located in relation to each other and the vehicle chassis to provide the combination of angles we call "wheel alignment." If the alignment angles are correct, the tires will not wear prematurely, and the vehicle will tend to steer straight ahead without wandering or needing constant correction. It will not lead left or right except as leading slightly by the crown of the road. Since road crown lead is generally to the right, engineers offset the alignment angles to counteract it. This offset, however, will cause a heavier lead if the car is driven in a lane having a left-hand crown. It is virtually impossible to eliminate lead in all circumstances, so a little lead on certain roads, or with certain loads or combinations of passengers, should not be cause for concern.

Suspension, steering, and wheel alignment should be checked whenever new tires are installed, and anytime uneven wear patterns are noticed (as you dutifully perform your weekly tire inspection and pressure check). Remember that misalignment may cause premature, uneven tire wear, but it will not cause vibration or shimmy. If you experience these conditions, even slightly, have your tires and wheels professionally inspected and balanced.

Charging and Cranking

Remember our lingo from chapter four? *Charging* is what must occur to replenish the power used from the battery by your car's many electrical demands. Charging is accomplished by the alternator, one of those bolt-on components at the front of the engine. The other component of the charging system is the voltage regulator, which keeps the charging rate under control by regulating the amount of electrical "pressure" delivered by the alternator. Nowadays, the voltage regulator is built into the alternator on nearly all vehicles.

Cranking is the mechanical spinning of the engine by the starter before it starts. In the old days (very old days!) the engine was actually "turned

HAND CRANKING

Davidson fondly remembers going to the barn to start up his 1957 MGA with a hand crank on cold New England mornings when the battery was not happy. Some motor oil warmed up on the family stove (much to his mother's dismay!) sometimes helped the situation. That car, which had a starter motor also, of course, may have been the last production car to come provided with a hand crank. Came in handy, too! Of course, that was a little 1500 c.c. four-cylinder engine of about seventy hp. Wouldn't want to try it on today's engines! He also remembers stuffing a carefully sized piece of cardboard in front of the radiator in winter to get the engine up to a reasonable operating temperature. Ah, those were the days!

over" with a crank, turned by the driver from the front of the car. Now the turning force is provided by the starter motor, which is engaged by a relay (called the *solenoid* on some models). The relay is operated by turning the ignition switch to the start position, which completes the circuit between the battery and the starter motor.

The functions of charging and starting are performed by entirely different components, but they have one major item in common: the battery. Your vehicle's battery is its electrical power source. On the modern car or truck, even a properly functioning charging system will not perform without a battery. The battery provides:

- 100 percent of the power needed to operate the starter motor during cranking

- Supplemental electric power to keep systems and accessories working properly at times of high demand

- Power to the alternator's field circuit to allow it to begin charging

- An electrical "cushion" of sorts to stabilize all of the vehicle's electrical and electronic systems.

As you can see, the battery is very important in any car or truck. It is perhaps the most critical single component for operating the systems that allow the engine to crank and run. Yet, as we mentioned, it is the component most likely to fail without giving much notice that anything is wrong. We talked about the battery and its usual life span of three years back in chapter three. Aside from actually testing the battery every few weeks, the only way to foresee impending battery failure is to observe changes in the operation of your car's other electrical systems, especially the cranking system. If the engine begins cranking more slowly than usual, or the lights seem to flicker a lot when the engine is running, get the battery tested, and *without delay!* The battery may be about to let you down—and batteries never choose a convenient time to do it. Some batteries are equipped with a "test eye" that appears green if the battery is adequately charged, red if it is discharged, and silver if the fluid is low.

Heating and Air Conditioning

When Roy first studied automotive air conditioning in 1971, about one car in five was equipped with air conditioning. Today, one would be hard-pressed to find one car or truck out of twenty without it. (Of course, that's not to say they all work!) An interesting by-product of

Figure 4.4—On many cars and trucks, the battery is easy to see and, therefore, easy to check periodically. Then there are cars like this one. To make matters worse, the "test eye" will likely be in a different location when the battery is replaced.

REFRIGERANT CONFUSION

If you have a car air conditioning system that uses Freon, and want—or must—change to an EPA-approved refrigerant, note that there is no such thing as a "drop-in" replacement refrigerant!

In the 1990s, increased awareness of the relationship between chemicals such as Freon and damage to the Earth's upper ozone layer caused the EPA to order a phase-out of Freon for use in automotive air conditioners. Its production ended altogether in December, 1996. Beginning in 1993, car makers began using the new R-134a. The performance of a system that is *designed* for R-134a rivals the performance of an old-fashioned Freon system. The primary design differences are that the evaporator and condenser are both slightly larger, the operating pressures are slightly higher, the hoses and seals are more leak-proof, and a different type of lubricating oil must be used.

Changing over to R-134a (called SUVA by DuPont, the company that makes it) involves specific procedures for different makes and models—some more extensive than others—and may require replacement of major system components. At the very least, the old oil will have to be flushed from the system, and the line fittings and receiver-dryer or accumulator and pressure switches will all have to be changed. On some models, the compressor and/or hoses may have to be changed as well.

Now that Freon (also a DuPont product) has gotten very expensive as supplies dwindle, and R-134a has gotten substantially cheaper (not to mention easier to get), it is worth considering making the change. Systems designed for Freon will cool best if Freon is used in them, but let's put this in perspective. Changing over to R-134a will raise the air conditioning outlet temperature by four degrees Fahrenheit in most cars; still, the average driver might not notice the difference, except on very hot, humid days. Even then, the car will still be cool.

Don't be tempted to put R-134a into a system that leaks. R-143a will leak out of an air conditioning system more easily than Freon, so if you have a small Freon leak, you will have a larger leak after changing to R-134a!

Quite a few "blend" refrigerants are on the market, some flammable, some claiming to seal system leaks, and some advertising "EPA approval." Understand that the EPA only tests refrigerants for their potential damage to air quality and the ozone layer, not for system performance, safety, or potential system damage! No blend refrigerants have been approved by auto makers, or endorsed by professional associations such as the Mobile Air Conditioning Society (MACS) or Society of Automotive Engineers (SAE). Some may be dangerous, and virtually any blend will cause recycling equipment to become contaminated. Reputable auto technicians will have no part of blends. You shouldn't either.

producing so many models intended for air conditioning is that cars and trucks are no longer being built with efficient fresh-air ventilation systems. It's tough to get good airflow across your feet and face without creating a veritable wind tunnel and enduring lots of noise. Maybe this is why many owners will neglect needed mechanical repairs and get the air conditioner fixed first!

Typical heating and air conditioning systems on today's cars consist of three basic systems that must operate together to create comfort in the passenger compartment. They are the engine cooling system, the refrigeration system, and the system controls.

The refrigeration system in your car operates like any other—your home air conditioner or refrigerator, for example. The principle is simple. Refrigerant is pumped through the system by the compressor. It enters the coils of the evaporator core in the passenger compartment at low pressure and low temperature. The blower fan circulates air over the evaporator core and heat is exchanged into the cold refrigerant passing through. The heat-laden refrigerant is transported out of the passenger compartment and under the hood, where it is raised to high pressure and high temperature. Then, it passes through the coils of the condenser, which looks like a second radiator mounted in front of the cooling sys-

NO PUMPING REQUIRED!

The true beauty of an ABS is that the steering still works, even if the driver is standing on the brake pedal. An ABS has the ability to pump the brake pedal much faster than a human can, and keep the wheels just short of locking up. Many modern systems are able to control braking to each wheel individually. Trust us: under any given circumstances, you cannot stop a car quicker than the ABS can! The most important thing to remember is to use the steering control you still have. We recommend taking your car to a safe place, such as a large (and empty!) parking lot, on a rainy day, and try out your ABS. Speed up to thirty mph or so, then do a few panic stops and try steering maneuvers at the same time. You'll be amazed! Also, this will give you an opportunity to see and feel the normal operation of your ABS so you won't be surprised later on.

tem radiator. The cooling system fan circulates air over the condenser and heat is removed from the refrigerant. This "refrigeration cycle" goes on continuously while the compressor is running.

Because the air conditioning system cools the passenger compartment by capturing heat from the air inside and depositing it in the air outside, there is a relationship between the engine's cooling system and the air conditioner's efficiency. If the engine runs hotter than normal, or the airflow across the radiator becomes blocked, the refrigerant cannot properly transfer its heat to the outside air. That means the evaporator cannot absorb as much heat and the result is poor cooling from the air conditioner. This is one more reason to keep an eye on the temperature gauge and the coolant level!

In addition to the refrigerant system, your car will have electronic components to control the operation of the compressor, blower fan, engine cooling fan, airflow outlets, and airflow temperature. We don't have enough space to explore all of the variations and operating modes, but we can spare a few lines for some more advice.

The best way to understand your car's heating and air conditioning system is to read the owner's manual (yes, here we go again!) and observe. Take a few minutes to learn what the system's features are and how to use them. You might be surprised to find that your system has some features you didn't expect! Then, observe how the system acts and performs in the various operating modes. By knowing how your system normally works you will be able to detect problems early and describe what is wrong.

Brake System

Every car and truck produced since the early days of the automotive industry has been equipped with two brake systems—a parking brake, sometimes called the "emergency brake," and service brakes. There are numerous brake designs and variations, but only two basic service brake types and two basic parking brake types have been commonly used since the 1960s.

The two basic service brake types are: disc/drum and four-wheel disc brake systems. Disc brakes have been used almost exclusively on the front wheels of cars and light trucks since the late 1960s. The reasons

Figure 4.5—This is a typical single-piston disc brake. Widely used worldwide since the mid-1960s, they are still the industry standard.

Figure 4.6A—Typical modern rear drum brake.

Figure 4.6B—Drum brake with the drum removed.

are simple: disc brake linings, called *pads*, wear longer than drum brake linings, and disc brakes provide more reliable controlled stopping with less tendency to grab, fade, or pull to one side. Since the front brakes are called upon to provide the majority of stopping power, up to 80 percent of braking on some front-wheel drive models, drivers universally appreciate these attributes.

About the only bad thing about disc brakes is they tend to be noisy. Disc brake pads tend to vibrate, whistle, and squeal somewhat because the pads are squeezed against the disc, called the *rotor*, which rotates between them. Some specific designs and lining materials are noisier (and more annoying) than others. On some makes and models, especially high-performance cars, smaller, less powerful disc brakes also are used for the rear wheels. On many others, the rear wheels use drum brakes, which have their own attributes.

Drum brake components are less expensive to replace than their disc brake counterparts. Drum brakes also apply more stopping force with less pedal effort than disc brakes. This feature makes them more effective for the parking brake, which is applied mechanically. Since the rear brakes do not work as hard as the front in stopping a vehicle, there is a tendency toward faster wear. The rear linings may outlast the disc pads by as much as 300 percent or even more! In addition, drum brakes are less prone to squeal or whistle while they work.

The parking brake is simply a mechanical means of applying the brakes on two wheels (almost always the rear wheels) to hold a vehicle stationary while it is parked on inclines. In many hilly areas, such as San Francisco, it is against the law not to use the parking brake when leaving your car parked. The usual design employs cables and mechanical linkage to apply the service brakes. A few of the "top end" models with four-wheel disc brakes, however, have small, mechanical drum brake units contained in the center of the disc brake rotors. These are truly "emergency brakes," because they are totally separate from the service brakes.

Antilock brake systems (ABS), originally called "anti-skid systems," have been around since the late-1960s. They have gained tremendous popularity in recent years. Like four-wheel drive, the ABS can be a valuable safety feature, but once again, it does not make your car, truck, or SUV invincible. It is also important to know that an ABS cannot prevent skids or wheel lock-up under all circumstances. An ABS is capable of only one thing: it prevents wheel lock-up during braking, so that the driver can maintain better steering control while stopping the car as efficiently as possible.

Most insurance companies still allow discounts for cars that are equipped with an ABS, but statistics have shown that it doesn't seem to reduce accidents or lower damage claims very much! We believe this to be caused by a general lack of training and awareness of the differences the ABS provides over conventional braking systems. Roy points out:

> Most of us still tend to freeze in a panic stop, without realizing that the ABS often gives us full steering control, even while we're standing on the brake pedal. You could drive your car straight toward a tree at thirty miles per hour, then stand on the brakes and still steer right around the tree, in a panic stop the whole time!

A good driver knows that pumping the brake pedal and not applying the brakes too hard are both helpful in controlling a car on slippery surfaces. That is exactly what the ABS does automatically, and more efficiently. When the brakes are applied and the sensors tell the ABS control module that one or more wheels are about to lock, the system takes over the braking, pumping the brakes on the locking wheel(s) about fifteen times per second and controlling the braking pressure. Sometimes, the ABS unit will actually push the brake pedal back up against the driver's foot. It is perfectly normal to see the ABS warning light illuminate, hear a loud clicking and clunking noise, and feel a rapid pulsation in the brake pedal while the system is operating. Remember that you are standing on the brakes, likely producing well over one thousand pounds per square inch of hydraulic pressure in the brake system. Your ABS has to overcome this pressure to take over, so there is a lot of force involved!

Traction control is a system that is only available on select models. It uses the same concept as the ABS, interrupting, pulsing, and controlling the brakes on individual wheels. But this system takes it one step further: now the control module knows when the vehicle is about to skid even when the brakes are not applied, and tries to help. The main difference is that more sensors are added to determine the force and direction of vehicle movement. Extensive engineering and testing is employed to know what is normal and what indicates an impending skid. A few really high-tech models even allow the traction control system to take over the accelerator as well as the brakes in their effort to keep drivers out of trouble. An expensive system? You bet it is, but it can be a real life saver. Roy describes his first test-drive in a Mercedes with traction control thus: "I couldn't believe it. The car was on a wet parking lot and I turned the steering from lock to lock at full throttle. I could not make the car break into a skid in any direction! The system offset everything I did wrong. Talk about idiot-proof!"

In the next chapter, we are going to look at your car's power train management system and go more deeply into the electronic controls we have been talking about. We'll see what makes those electronic sensors, valves, injectors, computers, and diagnostic links tick. Stay tuned, we're just getting warmed up!

YOUR CAR'S POWER TRAIN MANAGEMENT SYSTEMS

Americans are good with tools—it's almost a defining characteristic. The kinds of people who come here tend to be adventurous, resourceful, and ambitious, and some mechanical knowledge seems to go with those attributes. For nearly a century we did much of the world's manufacturing, and that was because we had an almost inexhaustible supply of skilled labor. Mechanical ability wasn't just the monopoly of factory workers and skilled technicians. Many of us can remember a time when it wasn't at all unusual for school teachers, accountants, and dentists to spend weekends working on their cars.

Times change, of course, and so do the tools. Though we still know how to hammer a nail or drill a hole, we seem to be leaving the mechanics to others. We are moving on to the digital age. Computers are widespread in our culture. Now we surf the Web—buying stuff, installing stuff, downloading stuff, playing sophisticated computer games and, on occasion, hacking our way into the Pentagon. We write most of the world's code, and we argue the

This lovely '92 Ford Explorer shows "00,407" on the odometer. The driver doesn't know how many times the zeros have "lined up," but since it is a business vehicle, we suspect this is not the first "rollover"!

merits of PCs and Macs the way we used to argue Fords and Chevys—or maybe Fords and Studebakers.

As we have mentioned, cars, too, have made the plunge into the digital age. This chapter goes into the computerization of the automobile in some detail. Maybe you feel like skipping this chapter, or giving it a quick skim. That's okay. There's enough good knowledge in this book, we hope, to make the rest worthwhile. Yet many readers out there will be quite at home with this discussion of the car's operating system, and it will certainly help create a more comprehensive and essential understanding of how the car actually works. If you enjoy learning "what makes it tick," this chapter is for you! Consider it another interesting application of cybernetics. Norbert Wiener, the thinker who coined the term *cybernetics,* would certainly understand it that way.

Actually, cybernetics is a good way to understand how the modern car's electronic systems work. The essence of cybernetics is control. The term is derived from the ancient Greek word *kybernetes,* which means helmsman, the sailor who controls the boat. The captain tells him the course, but to stay on it he must constantly adjust the tiller or wheel, taking into account his wake, the currents, the tides, the wind and the set of the sails, and their combined effect upon the boat's course. Think of a power train microprocessor as a helmsman in your car. What the microprocessor must do is take into account a whole range of variable signals from the various modules of your car's systems, and use that information to adjust your car's mechanical systems.

Electronics have become the vital, common thread that ties the systems and controls of your car together and keeps them functioning at peak efficiency. Although we will mention some accessories, such as air conditioning and power steering, we will concentrate our efforts on the engine, transmission, and emission controls: the team known as the power train management system.

DO YOU KNOW WHY YOUR RADIO VOLUME CHANGES?

Sometimes, you discover a feature on your new car that you did not even notice. (Your salesperson may not have known about it, either!) For example, if you should see a small lever peeking out from behind your radio's volume control, your car's audio system might have the capability to increase its volume to offset the increased road and engine noise as the vehicle speeds up. On many new models the amount of volume increase is adjusted by that little lever.

This is just one example of how the electronics of the modern car are interrelated. This little convenience requires involvement from the engine computer, vehicle speed sensor, and audio control module. Of course, the owner's manual is the place to find a full explanation of how to use this, and all of your new car's other nifty features.

The Power Train Management System

Electronic modules are used to control what previously were purely mechanical functions, as well as accessory systems, such as climate control, cruise control, and audio systems. Since many systems on today's cars and trucks are interrelated, the modules often communicate with one another. For example, the stereo volume may be programmed to increase as vehicle speed increases. The same vehicle speed signal may also be used by another module to help decide when to shift the automatic transmission.

A *control module* is a microprocessor unit—a computer—that controls or operates one or more specific components. The module may work independently, or it may use shared information with other interrelated systems or components. For example, the electronic brake control module (EBCM) operates the antilock braking system (ABS) by controlling hydraulic brake pressure to each wheel in response to the speed signals it receives. On many vehicles, some of the same information is used to operate the power train control module(PCM) and speedometer.

Back when microprocessors were first being used in cars, it wasn't unusual to find seven, eight, or nine control modules independently controlling various automotive functions. Today's cars mostly use no more than three control modules: the power train control module (PCM); the body control module (BCM); and the electronic brake control module (EBCM).

That doesn't mean that on-board electronics have gotten simpler. The Mitsubishi 3000 GT and Diamante have sensors mounted in the headliner to tell the automatic air conditioner the temperature inside the car and another on each strut to monitor the position of the suspension at the four wheels. There are still other modules: the 2001 Cadillac DeVille has twenty-four "class II" electronic modules. "Class II" means that these modules send and receive communications to each other, to the PCM, and to a technician's test equipment. The 2001 Buick Park Avenue has sixteen such modules. The number of electronic devices used is determined by the number of functions and accessories with which a car is equipped with. These devices communicate critical information to each other; their functions are often combined to produce a specific result from one of the main computer control modules.

Typical Power Train Control Signals

Signal	Function	Used for
Manifold absolute pressure sensor (MAP)	Information, engine load	Fuel mixture, ignition timing, transmission shifts, emission control devices.
Barometric pressure sensor (BARO)	Information, altitude, barometric pressure	Fuel mixture, ignition timing, transmission shifts.
Throttle position sensor (TPS)	Information, exact throttle opening	Fuel mixture, ignition timing, transmission shifts, EGR valve operation.
Intake air flow (MAF)	Measures speed/density of intake air	Fuel mixture, idle speed, transmission shifts, emission control devices.
EGR valve position (EVP)	Feedback, tracks current EGR position (% open)	Fuel mixture, ignition timing, EGR valve operation.
Vehicle speed sensor (VSS)	Information, measures vehicle road speed	Speedometer, cruise control, transmission shifts, ABS, traction control, EGR valve operation.
Wheel speed sensors	Information and feedback, measures speed of each wheel	ABS, traction control operation and self-diagnosis.
A/C request	Feedback, indicates whether A/C is turned on	Fuel mixture, idle speed, A/C self-diagnosis.
A/C on	Feedback, indicates whether A/C compressor is engaged.	Fuel mixture, idle speed, A/C self-diagnosis.
Transmission torque converter clutch on	Feedback, indicates whether transmission converter should be locked	Fuel mixture, ignition timing, transmission shifts, transmission self-diagnosis
Transmission 1, 2, 3, 4 gear switches	Feedback, indicates whether shift commands were successful	Fuel mixture, ignition timing, transmission shifts, transmission self-diagnosis
Engine rpm	Information, measures engine speed	Fuel mixture, ignition timing, idle speed
Crankshaft position sensor	Information, tracks position of engine crankshaft	Timing of fuel injections, ignition timing
Camshaft position sensor	Information, tracks position of valve train components.	Timing of fuel injections, ignition timing.
Power steering pressure switch	Information, reports increased engine load from parking maneuvers	Idle speed increase.
Coolant temperature	Information, measures operating temperature	Fuel mixture, ignition timing, idle speed, EGR valve operation, emission control devices.
Intake air temperature	Information, measures temperature of air entering engine	Fuel mixture, ignition timing, EGR valve operation, idle speed.
Oxygen sensor(s)	Feedback, measures actual fuel combustion efficiency, reports rich or lean mixture in exhaust	Fuel mixture, ignition timing, EGR valve operation, emission control devices.

Figure 5.1— Powertrain Management Signals

What Computers Control and How

Signals *(inputs)* provide the information that is used by a computer in deciding exactly what action to take. Signals have three basic functions:

1. To supply precise information that will be used, along with information provided by other signals, by the PCM to calculate the next action. For example, the computer uses information about the exact position of the crankshaft pistons to send the next spark to a spark plug or to inject the next charge of fuel into the intake system.

2. To let the computer know what is going on in various systems and components. For example, on-off switches are used to tell the PCM what gear the transmission is in, whether the air conditioning compressor is running, and whether the power steering is being "loaded" during parking maneuvers.

3. To track a computer-controlled command. For example, if the PCM has commanded the EGR valve to open 30 percent, a second signal may be used to measure how far the valve actually opened. So, in effect, the actual valve position is "relayed" back to the computer as it operates. This type of information is referred to as *feedback*.

The following chart lists the most common signals related to the PCM and how they are used. You will see how the operation of many of the various controls and systems are interrelated and how a single function requires the use of many signals.

Computers use the array of information gathered from the signals to make decisions that result in actions or commands, known as *outputs*. For example, a coolant temperature signal that moves out of the computer's pre-programmed normal range for a pre-programmed period of time will cause the malfunction indicator lamp (MIL) to be illuminated, giving the driver a message, such as "check engine" or "service engine soon."

Inside the computer there are four main sections of activity:

- Input conditioners
- Memory
- Microprocessor
- Output drivers

Input Conditioners

The signals sent directly from the sensors are often not adequate for the computer to use directly. Computers cannot understand variable analog signals; they must be translated into digital data to be understood. Input conditioners receive the analog inputs and translate them into bits and bytes for the microcomputer. Some signals need to be made stronger by an amplifier. Others will need to be "cleaned" by a buffer for better accuracy.

Memory

There are several kinds of memory required by the microcomputer. Virtually all of the pre-programmed information the computer must work with is stored in the read-only memory (ROM), like the information on your hard drive. Think of the ROM as the computer's reference library, where it can "look up" the appropriate response for a given set of signal values. It is the basic set of instructions on how information will be handled.

There is also programmable ROM (PROM), which stores data that must be custom-tailored for each vehicle, taking into account such things as weight, engine, transmission, axle ratio, and accessories. ROM and PROM are permanently "burned in," which means that they will not change if the computer loses power or is removed from the vehicle.

To make it possible to modify the PROM as changes in programming are needed, three variations are used: erasable PROM (EPROM), electronically erasable PROM (EEPROM), or flash PROM. The information contained in these erasable PROMs is changeable using special reprogramming procedures. Once changed, their programs are permanent until reprogrammed again. The technology is similar to that used in a digital camera, which stores pictures on a permanent memory card, but allows the images to be cleared and replaced with new pictures.

What this does is make it possible for a dealer's service department to make approved changes—updates—to correct a design flaw or other problem that was not apparent before the vehicle was built. The "fix" may be as simple as plugging into the car's diagnostic link and updating the PROM with new data. The process is similar to fixing bugs in computer applications.

Unlike the ROM information, random access memory (RAM) is active and volatile, just like the RAM in your personal computer. Inputs received by the computer are changing more or less continuously as the car is driven, which, in turn, changes the specific actions the computer takes. The inputs can be put in, taken out, updated, and rearranged at random. Acting on signals entered into RAM, such as vehicle speed, engine speed, load, and throttle position, the computer will adjust the fuel delivery, ignition timing, EGR position, and various other functions as instructed by ROM and PROM. The RAM is essentially a notepad, where the computer jots down information and calculations, with much of it always being erased and changed.

The third type of memory is a variation of RAM. Conventional RAM is lost when the key is switched off, so the computer starts with a clean notepad each time the car is started. This other type of memory, commonly called "keep alive memory" (KAM) can survive as long as the car's battery remains connected and adequately charged.

The concept is the same as electronically tuned radios, which retain the selected preset stations until the battery is disconnected. In the engine computer, KAM allows each vehicle's fuel injection system to be fine-tuned to compensate for such variables as a small intake air leak, high altitude operation, and minor sensor signal inaccuracies. Also, KAM can help to maximize performance and economy according to the way the car is driven. The computer can "remember" the adaptive adjustments it has made until battery power is lost.

Microprocessor

The inputs in RAM must be coordinated with the instructions in ROM, and this occurs in the microprocessor. Here, based on the data in memory, split-second decisions are made to properly execute all functions and operate all components within the computer's control. Once an appropriate action is known, the microprocessor sends the operating instruction to the final activity center, the output drivers, similar in function to the various drivers in your personal computer.

Output Drivers

The output drivers are largely comprised of tiny transistors. These transistors function as "solid-state" switches, meaning that they have no moving parts. The microprocessor uses these switching transistors in various combinations of "off" and "on" to perform its control functions. Many operations are accomplished directly by the computer, but the operation of heavy-duty circuits requires the use of relays, because the computer's tiny output drivers cannot tolerate high amounts of current flowing through them. A relay is a two-stage switch that uses a small control circuit to operate a large, higher-current circuit. Depending on the application, a relay may be mechanical, with moving parts, or may be solid-state.

The Five Signal Types

Any piece of information that travels from one electronic unit to another is called a signal. Signals are sometimes referred to as the building blocks of intelligent electronic communication between sensors, control units, and other devices in a system. The computer must translate all of the signals it uses into a common code of electrical values in order to process the information it receives and generate the appropriate output commands for the components it controls. Every signal is decoded, then used by the computer for decision making. There are five primary signal types used in all of this high-tech interaction. They are:

- direct-current signals
- alternating-current signals
- frequency-modulated signals
- pulse-width modulated signals
- serial-data signals

These five signal types come in two basic forms, analog and digital.

Direct-current (DC) signals are characterized by a current flow in one direction through a circuit. They are either simple on-off signals, or variable "analog" signals. They are generally used for power supplies sent by the computer to sensors, such as battery voltage or a lower, reference voltage generated by the PCM; signals from analog devices such as temperature sensors, pressure sensors, throttle and EGR position sensors, and oxygen sensors; and switches to indicate the status of components as off or on.

ANALOG AND DIGITAL SIGNALS

The terms **analog** and **digital** describe two different ways of presenting information. A simple example can be found in the wristwatch. Most wristwatches today are still analog devices. One looks at the watch and notes the position of the hands to tell the time. There are digital wristwatches also, which present the time as a series of counter numbers. Both can be accurate. The analog watch has an advantage in that it can be read at a glance, even from some distance, and can give a quick relative sense of the time, and where you are in the progress of the day. The digital watch has an advantage in that it tells time to the nearest second and resolves time into pure numbers. Still, many people feel that analog is better for watches. It's not for microprocessors, though.

Up until twenty years ago or so, most information was analog, even in high technology. Much of it still is, though most communications technologies are headed for digital. Microprocessors like everything digital, so analog information must be converted to digital information so the microprocessor can use it. Yet most of a car's sensors, the devices that directly measure outputs from power train units, deliver their information in analog form. An analog signal is one that varies on a more or less continuing basis, and is controlled by a regulating device that responds to whatever is being measured. We can think of these signal control devices much like faucets or shut-off valves in a water supply line. As the faucet is opened or closed, the amount of water flowing through the pipe is increased or decreased.

The signal often begins with a power supply to the sensor, usually five volts, that is accurately regulated by the PCM and sent out to the sensor. At the sensor, this reference voltage is reduced to indicate its measurement and then that reduced voltage is sent back to the computer. For example, at the signal wire from the throttle position sensor, a closed throttle of an idling engine would have a voltage reading near zero. The same sensor reading would be near five volts at wide-open throttle. In between, the voltage would vary continuously as the throttle is moved, increasing and decreasing in direct response to the driver's accelerator movements.

Some analog signal devices generate their own voltage as the conditions they are measuring change. These devices do not require the use of a power supply to operate. One example of this is the exhaust oxygen sensor, which generates voltage chemically (somewhat like a dry cell flashlight battery) in response to the amount of oxygen in the exhaust stream. The less oxygen, the higher the voltage; more oxygen, lower voltage.

Digital signals are characterized by a distinct on-off or high-low cycling of their voltage. Digital signal devices generally require the use of a reference voltage, but instead of varying the voltage returning to the computer, they indicate their measurements by varying the relationship of "high voltage" to "low voltage," or "off" to "on," over time. With a digital signal, the voltage constantly varies between the same high and low (or on and off) values, just as though a switch were being turned off and on. What varies is how long the voltage stays high in relation to low (or on in relation to off) or how many times per second it changes.

STOP

WIRELESS INTERNET DEVICES

An incredible array of new wireless Internet devices are flooding the market, literally at the speed of the Internet itself. These units include specially equipped cell phones, interactive pagers, and personal digital assistants (PDAs), such as the popular Palm Pilots. Wireless Internet access devices have opened the door for an unbelievable variety of services, games, music, and other conveniences—without your having to wire the unit into your car or be connected to an Internet service provider. They perform virtually all of the functions available from your home computer, including offering safety and security.

Trouble Codes
1995 Dodge Dakota

Code	Definition	MIL Light On?
11	No crank position signal	Yes
12	Battery disconnect/power loss	No
13	No change in MAP from start to run	Yes
14	MAP sensor voltage too low	Yes
15	No vehicle speed sensor signal	Yes
17	Engine is cold too long	No
21	Oxygen sensor stays at center or shorted to voltage	Yes
22	Coolant temp. sensor voltage too high or too low	Yes
23	Intake air temp. sensor voltage too high or too low	Yes
24	Throttle position sensor voltage too high or too low	Yes
25	Idle air control motor circuits	Yes
27	Fuel injector control circuit (any cylinder)	No
31	EVAP solenoid circuit	Yes
32	EGR solenoid circuit or EGR system failure	Yes
33	A/C Clutch relay circuit	No
34	Cruise control solenoid circuit or switch	No
35	Radiator fan relay control circuits	No
41	Alternator field not switching properly	Yes
42	Auto shut down relay control circuit	Yes
44	Battery temp. sensor out of limits	No
46	Charging system voltage too high	Yes
47	Charging system voltage too low	Yes
51	Oxygen sensor signal stays lean	Yes
52	Oxygen sensor signal stays rich	Yes
53	Internal PCM failure	No
55	Completion of diagnostic codes display	No
62	Service reminder indicator miles not stored	No
63	PCM EEPROM writing (programming)failure	No

Figure 5.2—Trouble Codes

Alternating-current (AC) signals have a characteristic current flow that rises above and falls below a specific base level, usually zero, in repetitive cycles. They are most often used for "timing" signals. AC signals are found in such components as:

• ABS wheel speed sensors, to signal a wheel that is on the verge of locking up

• Vehicle speed sensors, to help the computer decide when to shift the transmission and make changes in the cruise control operation

• Knock sensors, which generate their own AC voltage at a particular sound frequency so the computer can "hear" a spark knock electronically

Frequency modulated signals are generated using digital sensors that vary the number of times per second (the frequency of) their voltage switches between high and low. Typically, frequency-varying signals are used for:

Figure 5.3—This typical pre-1996 GM diagnostic link will cause the PCM to flash trouble codes using the "Service Engine Soon" lamp when the test terminals are jumped with a wire as shown.

- Digital mass airflow sensors, which measure the amount and density of air coming into the engine
- Digital MAP sensors, which measure engine load
- Crankshaft and camshaft position sensors, which tell the computer the exact location of the engine's rotating parts

Pulse-width modulated signals are more often used by the computer as output commands to operate electronic devices rather than being used as input signals. The variable in this type of signal is the percentage of time the voltage is held high or on, compared to the low or off time. For example, more fuel can be delivered to a cylinder by holding an electronic fuel injector open for a longer fraction of a second each time the intake valve opens during the engine cycle. If the injector is held open for less time per engine cycle, less fuel will be delivered. This type of signal is commonly used for the ignition coil; electronic spark (ignition) timing; EGR control solenoids; evaporative emission control solenoids; fuel injectors; and idle speed control motors or solenoids.

Serial-data, or "multiplexed," signals are produced by the PCM, BCM, EBCM, and other control modules to communicate with each other and with the auto technician's diagnostic equipment if the vehicle is equipped with self-diagnostic capability. *Multiplexing* means that more than one signal can be carried by the same wire or circuit at the same time. This technique of multiplexing signals has been used by manufacturers of stereophonic audio equipment for over forty years. It became popular on automobiles in the early 1980s. We'll have more information about self-diagnostics a little later in this chapter.

"Drive-by-Wire" Systems

As cars and trucks continue to get more sophisticated, electronic controls are being used more and more. The latest electronics are taking over functions that were once entirely mechanical. As we go to press, several late models have no linkage or cable connecting the accelerator to the engine's throttle. Further, electronic "active" suspension systems that not only sense oncoming bumps but actually move the car's wheels up and down to offset them, are very close to mass production.

Although some weatherproofing issues still challenge engineers, totally electrically operated steering is now a reality, with no mechanical parts

Figure 5.4—This is a modern scan tool, used to read the computer's data stream and stored trouble codes. All of the cables, often referred to as "spaghetti" by technicians, were necessary to read data on pre-1996 cars.

between the steering wheel and the steering gearbox. Add to this the ever-growing number of electronic accessories and comfort features, such as navigational systems that track a vehicle by satellite and computers that automatically call for roadside service when an impending breakdown is sensed. One can easily see just how far the businesses of automotive design and repair have evolved since the days of the "muscle cars," back in the 1960s. What's in the future? Well, at some point down the road look for cars that drive themselves. Don't laugh! They are working on it.

So many new electronic features are on the horizon—and already here—that alternator overheating and noise, as well as reduced fuel economy from the huge electrical demands, are serious concerns. To address these problems, a worldwide consortium of automotive engineers has developed liquid-cooled alternators and the forty-two–volt electrical system, soon to be the new industry standard. Look for it to appear soon in a showroom near you!

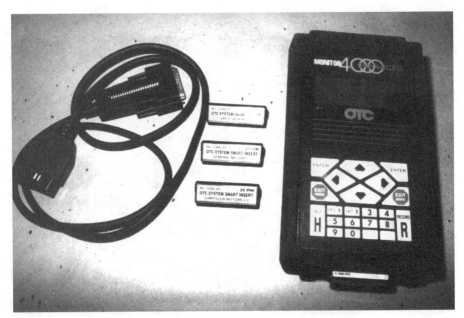

Figure 5.5—For 1996 and later models, these four adapters are all that are needed in order for the scan tool to read data from the car's computer.

Self-Diagnostic Systems and How They Are Used

Self-diagnostic systems began to become common in the early 1980s, as engineers realized that many hours of service time were being spent by technicians (the then-new name for mechanics) trying to figure out which erroneous signal caused the engine to cough up black smoke. Perhaps the most frustrating challenge of all was found in troubleshooting intermittent, "phantom" problems.

Designers and programmers rose to the challenge by equipping automotive computers with the capability to check their signals, both inputs and feedback, against known, "normal" values stored in the ROM. If a signal falls outside of an expected value with a given set of conditions, or if a component fails to respond to an operating command, the computer is instructed to turn on the MIL to warn the driver that there is a problem, and to store a trouble code that the technician can retrieve to see which signal was off-target. To avoid false alarms from a momentary glitch, the signal or response must remain abnormal for a predetermined time period, usually about two minutes.

This chart lists the trouble codes and their definitions for a 1995 Dodge Dakota. A list of trouble codes and the diagnostic procedures used for each are found in the service manual for the specific vehicle. Trouble codes often are now universal, and may change from one model year o the next. Each trouble code corresponds to a specific diagnostic troubleshooting procedure found in the service manual.

The presence of a trouble code does not tell the technician what to fix, but it does tell her or him which area to concentrate in while troubleshooting. The manufacturer has established a diagnostic procedure for each trouble code, charting a prescribed order for testing components and circuits until the problem is located. On some models, trouble codes may be read without special test equipment. On many Chrysler products, simply turning the ignition key on and off three times and leaving it on the third time will cause the PCM to flash out trouble codes for the technician. On many General Motors vehicles and some Asian imports, inserting a jumper wire or paper clip between two test terminals will retrieve the codes. On many Nissan models, turning a screw on the side of the PCM triggers the codes. Others, including virtually all 1996 and later models, require the use of a tester, such as a *scan tool* to retrieve the codes.

Most cars produced since the late 1980s can communicate with the scan tool to allow the data stream of inputs, outputs, and feedback to be monitored while the engine is running. The scan tool is plugged into the car's diagnostic link, located under the dash or in the engine compartment, and the computer's conversation with the technician begins. Like other technology, self-diagnostic systems continue to evolve. Each year finds more detailed trouble codes and enhanced information to guide the technician to the root cause of a malfunction.

Before the 1996 model year, most diagnostic equipment, and the codes they read, was proprietary—the information could only be read using a particular piece of test equipment available only through the manufacturer. Beginning with the 1996 model year, the EPA mandated that all cars and light trucks sold in the United States must have a standard diagnostic link and universal computer language to allow the reading and monitoring of nearly all engine management functions using a single, standard scan tool. This information could no longer be proprietary. This new universal self-diagnostic system is called on-board diagnostics, second generation, or OBDII for short.

Unfortunately, the legislation stopped short of specifying a standard access protocol. Because of this, there are four different diagnostic protocols used on OBDII vehicles. Although the diagnostic connector is the same size and shape on every model, four special adapters are need to make the computer "talk" to the scan tool! That's one more equipment investment the repair shop of today must make, but it is still better than a few years ago when every make had its own special connector and required unique "spaghetti" to connect a scan tool.

In the next chapter, we will take a look at how you can use your understanding of this high-tech machine of yours to guide you to a resolution when things go wrong. You may be surprised at how much you can still check yourself, and no, you don't have to buy a scan tool or a lab scope! We'll help you and your car stay damage-free and you'll learn a few fun things, too!

CHAPTER 6

WHEN THINGS GO WRONG

But Mousie, thou are no thy-lane,
In proving foresight may be vain:
The best laid schemes o' Mice an' Men,
Gang aft agley,
 An' lea'e us nought but grief an' pain,
For promis'd joy!

> — *Robert Burns,* To a Mouse

Well, things go wrong even with the most beautifully maintained vehicles—and other things besides. Meanwhile, you are late for work or, worse, stuck on the Interstate, tearing your hair out. It happens, and it's best to be philosophical about it. There's a bumper sticker on the back of an old pickup truck somewhere that describes the situation even more colorfully. Perhaps you've seen it. Maybe it will help.

As far as the electronic control systems we discussed in the last chapter are concerned, the good news is that many failures in the electronic control systems will not leave you on the roadside. Most recent car models are equipped with a back-up program, common-

THE NORTHSTAR ENGINE

The later the vehicle model, the more sophisticated the back-up system is likely to be. For example, the NorthStar V-8 engines used in recent Cadillac models and the Oldsmobile Aurora are programmed so that the engine can continue to be driven without damage if the cooling system fails! The PCM cuts off power to part of the cylinders, and then rotates which cylinders are working to reduce and equalize the operating temperature throughout the engine. Horsepower and performance are drastically reduced, to be sure, but the car can continue to run without immediate major damage. Continuing to drive with an overheated engine on an older Cadillac will cause the engine to seize from the increased heat and friction and literally melt internally. The likely result will be replacement of the entire engine—and a hefty repair bill, to say the least!

ly called a "limp-home mode," to try to help you reach your destination, or at least the next available service facility.

Most problems that actually cause a car or truck to become disabled on the roadside are failures in the basic systems, not the electronics. While we don't wish to belabor the point, many of these failures are the result of improper or inadequate maintenance. Surprisingly, there are still a number of things that can be tested, repaired, and maintained at home, even on the electronic, computerized animal we have been describing. If you have a fair amount of mechanical aptitude and enjoy taking things apart—and putting them back together again!—you can still experience the joys and frustrations of working on your own car.

To begin with, the tools and techniques that must be used to diagnose problems are different, largely because the sensitive electronics that seem to be everywhere must be protected. We are not talking about investing in an expensive scan tool or lab scope. For the most part, basic components are still serviced and replaced using the same basic tools as always. There are, however, a few dos and don'ts that are necessary to work efficiently on today's technology and keep out of trouble at the same time.

The Tools to Use—And Not to Use!

Beyond the basic set of wrenches and screwdrivers—with an oil filter removal tool thrown in for good measure—the list of tools needed to do what can be done at home or on the roadside is neither long nor expensive. You will need:

- A three-foot length of flexible rubber or plastic vacuum hose. That's right, vacuum hose. It is better than a stethoscope for listening! Just bear with us. We'll show you how!

- A high-impedance, digital multimeter. A multimeter is one that can measure volts, amps, and ohms (we'll get to the actual measurements later). "High impedance" means that the meter takes a very accurate reading without allowing much electrical current to flow through it. This prevents overloading and protects the tiny circuits in the PCM and other electronics, even if the meter should be connected wrong. Surprisingly, accurate meters are quite inexpensive (about $60 as we go to press) and are available at most auto parts stores and electronics shops. When shopping for a multimeter, just be sure that the input impedance rating is at least 10,000,000 ohms (10 megaohms) per volt. That meter will do no harm! You don't need a $300 meter unless

SAFETY GLASSES

Safety glasses should be worn anytime you are doing anything under the hood or under the car. Battery acid, refrigerant, and other chemicals found in the modern car or truck can cause serious eye damage. It is also a good idea to wear gloves to help protect your hands from chemicals and hot surfaces. Many technicians now wear surgical gloves whenever they work on or around the engine.

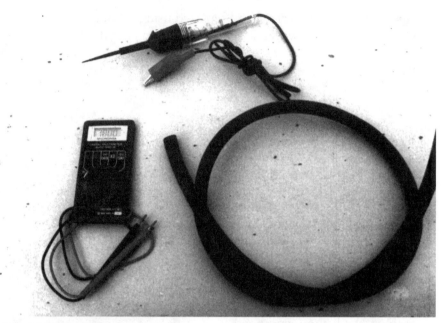

Figure 6.1—These simple tools can be used to test for many problems, even on high-tech electronic systems.

you are a professional technician. The difference is mainly durability. *Heavy-duty meters without high impedance ratings must never be used except on the battery and cranking system.*

- A twelve-volt test light, the kind with a sharp metal probe at one end and a fairly long wire with a clip at the other—the longer the better. These also are readily available at local auto parts stores. *Proceed with extreme caution here: these lights are to be used only to check for power to heavy circuits and components.* They *can* overload and damage electronics, and may damage wire insulation as well. Do not succumb to the temptation of testing wires and components at random. You may create a very expensive mess! We'll have some reliable guidelines for you just a bit later in the chapter.

- A keep-alive unit to hold all of your car's electronic memory settings when you need to disconnect the battery. We recommend the type that uses a nine-volt battery and plugs into the cigarette lighter or power outlet in the dash. If your lighter only works with the key on, you will need an adapter with clamps to hook the unit to the battery cables.

- Safety glasses.

- A battery cleaning brush.

- Jack stands must always be used if you will be working under the car for any reason—even for "just"changing the oil! We recommend using them for any activity that requires wheel removal, such as replacing brake pads or changing a tire, even when you will not be under the car. It's not uncommon for a car to slip off of a jack and be damaged—even when the person not using jack stands is a professional (a lazy professional, that is). A car slipping off the jack is also a hazard to humans working under or near the car. The doctors have a word for it—mechanical strangulation.

Remember that it is important to use the right tool for the right job, no matter what you are testing or repairing. For example, pliers are very handy to use for gripping and holding small objects, but they are not intended for turning or holding nuts or bolts. Using them in that manner is almost sure to cause damage to the bolt head and/or personal injury—skinned knuckles, bashed elbows, and the like. The same goes for "universal" tools such as adjustable wrenches. They are not designed to withstand the high torque needed to tighten and loosen fittings, and it is nearly impossible to use them effectively in tight quarters. Not much free space is found under the hood nowadays.

It is better to stop working and get the tool you need than to break something (on you or the car) trying to use a substitute. Of course, you must weigh the decision on which tools to buy against the cost of simply having the job done. Remember to consider whether or not you will ever use that tool again. Buying a special socket to install an oil sending unit on your car may not make sense, since it is very unlikely that the car will need another one during the course of its life. But a special socket used to replace brake pads will likely be used every 15,000 to 30,000 miles, so it will easily pay for itself. You are going to be driving that car for 300,000 miles, remember?

THE TROUBLESHOOTING PROCESS

We don't know who the first systematic troubleshooter was, but the term has certainly been in use throughout much of the twentieth century. We suspect that the process is older, however. As machine design became more sophisticated throughout the period of the Industrial Revolution, breakdowns proliferated. At first, diagnosis was rough and ready. As one early-nineteenth-century American designer of woodworking machinery used to say, "Let's start her up and see why she doesn't work!"

In time, however, engineers learned how to rationalize the process, and the troubleshooting process as we know it today is the result. It is used in every branch of engineering and in service shops of every kind throughout the world. Essentially, it is a process of binary elimination, confining the problem to a smaller and smaller area of the system. Most physical problems lend themselves to solution by the process, but not all. Intermittent problems, problems that cannot be replicated on demand, are often resistant.

Still, it is the best system we know of for resolving problems in the physical realm. Those interested in learning more about this should consult Steve Litt's great Web sight, the Universal Troubleshooting Process, at: http://www.troubleshooters.com/tuni.htm.

Basic Diagnostic Procedures for Any Car

When diagnosing a problem look first at the basic, obvious causes and work down to pinpoint testing only as needed. The fact is that 80 percent of all breakdowns are not related to electronic components or controls. If you experience a breakdown, or if something isn't working correctly, first think back to your knowledge of how the affected system or component is supposed to work. Remember the observations you made after reading the previous chapters. Take time to go back and review the appropriate passages in chapters four and five. If you have been following our suggestions, then you know what your engine normally sounds like when it cranks, starts, and idles. If it won't run this morning, what sounds or feels different? That is the same starting point that a technician will use to troubleshoot any complaint. Actually, step one in almost any diagnostic procedure is to verify the problem, if possible.

Once you have determined what seems to be abnormal, you can use these symptoms to guide you to the problem. The following are logical diagnostic test procedures that will help to identify which system or component is at fault and troubleshoot problems on any car. We'll begin with testing for power to electrical components, then discuss some common engine problems.

In the case of a heavy-duty electrical component or light that will not work, it is generally okay to use your test light. Begin by verifying that the test light is working. To do this, simply connect the test light across the battery terminals. Attach the clip to one terminal and touch the probe to the other. It doesn't really matter which terminal you use for the clip, because the test light will light either way. Any time your test light illuminates in this manner, it means that you have full battery power between the clip and the probe. Simple enough, right?

Ready to make some tests? Great! Let's begin by identifying the battery's positive and negative terminals and take a few practice readings. To help identify the negative and positive terminals, most auto makers have the positive terminal cover or cable colored red, while the negative is black. If the battery has post-type terminals on the top, the positive post is always slightly larger in diameter than the negative. When you think you have the positive terminal identified, attach the test light clip to it. Touch the probe to the other battery terminal to make sure the test light is working. It should light. Now, move the probe around and touch it to various objects, except wires and connectors, under the hood. Do not push the sharp tip into insulation or painted surfaces. When you touch most unpainted metal objects, the test light should light. If not, you have attached the clip to the negative terminal.

Any surface that lights the test light (when the test light is clipped to the positive terminal) is connected to the negative battery terminal, either directly or indirectly. All of these negative surfaces are called *grounded locations,* or simply *grounds.* Because all of these grounds are sort of

Open Circuit Voltage Readings	
Open Circuit Voltage Reading	**State of Charge, 12 Volt Battery**
12.5 - 12.7 volts	100%
12.4 volts	75%
12.2 volts	50%
12.0 volts	25%
11.8 volts or less	0%*

*Note that a battery with one dead cell will read approximately 10 volts, even when fully charged. Each cell produces 2.1 volts at full charge.

Source: Battery Council International

Figure 6.2— Open Circuit Voltage Readings

like one big negative terminal, the automobile is said to have a *single wire, common ground electrical system*. The reference to "single wire" is because simple circuits, such as lights, can be wired using only one positive wire going to the component. The ground path for the circuit is completed by mounting the component so that it touches a negative surface on the engine or body.

Another way to determine which battery terminal is which is to get out your multimeter. Begin by reading the directions to familiarize yourself with its operation. Set it to read DC volts in a range of zero to twenty, or at least over twelve, and touch the two test leads to the battery terminals. If the voltage reading has a minus sign, the red test lead is on the negative terminal. If not, the red lead is on positive.

Checking Out the Systems and the Symptoms

Cranking and Charging Troubles

The battery should always be considered the prime suspect when the engine is not cranking properly or when electronic systems and accessories are not functioning correctly. As we discussed earlier in the book, the battery is prone to failure after three years or so, and there is no sound or warning light to tell you when it is getting weak. But using your multimeter, you can check your battery any time you like. Reading the battery voltage will tell you its state of charge, and whether it has adequate power to crank your engine reliably. In fact, you can check out your battery, cranking system, and charging system, all in about ten minutes with a few simple tests.

The first battery test is called *open circuit voltage*. This measurement indicates the battery's state of charge and can also lead you to defects, such as one cell that has failed. Since "open circuit" means that the circuit is not operating under any load, step one is to make sure that all systems and accessories are turned off, including the cell phone and under-hood light. To turn off the light under the hood, simply unplug the connector using a small screwdriver to gently release the holding clip, or remove the bulb. Be careful, as the bulb can get very hot.

Once everything is turned off, connect your multimeter across the battery terminals with the range set above 12 on DC volts, and read the voltage. Although we refer to a 12-volt electrical system, each battery cell actually produces 2.1 volts and there are six cells, which adds up to 12.6 volts. Surprisingly, the cell voltage changes very little as the battery becomes discharged. A battery that is only 25 percent charged still reads 12 volts. A voltage of less than 12 indicates an internal battery problem. If the voltage is between 10 and 10.5, it is a sure sign of a "dead" cell, one that is no longer producing any power. We should mention that it is possible for the battery voltage to read above 12.6 if the engine has been running recently. This is called a *surface charge* and it must be removed for accurate testing. To remove the surface charge, simply turn on the headlights for fifteen seconds, then read the voltage again. It will be accurate this time!

Refer to the following chart to measure the state of charge for any 12-volt automotive battery.

If the battery voltage indicates a good state of charge, 75 percent or more, you can do a second test to see if it still has adequate cranking power for your engine. Leave the multimeter connected to the battery, and position it so that you can see the reading from the driver's seat. Now, crank the engine and observe the reading before the engine starts (if you can—you may need another person to assist you in this). The

CLEAN THE BATTERY, LOCK THE DOORS!

An interesting little quirk is programmed into the anti-theft system of some Land Rover vehicles. If the battery has been disconnected or has lost all power, the computer thinks the car is being stolen when the battery is hooked up again or jumper cables are connected. It responds by immediately locking all of the doors, in a "super lock" mode. Conventional unlocking methods, such as lockout tools, cannot be used to gain entry and there is no way to override the system from under the hood. The only way to get in that we know of is to obtain a new key from the Land Rover dealer or have one cut by a locksmith. In either case, you will almost certainly have to identify yourself as the rightful owner to get the new key. The obvious solution is to have the keys in your pocket before you lay a hand on the battery. It's good advice with any vehicle!

voltage should remain over 9.5. If not, the battery is getting weak, or it is possible that the starter is using too much current. It is time to have the battery professionally load-tested to confirm the diagnosis. Your local auto parts store will most likely provide this service at no charge. If the battery turns out to be the culprit, replace it now—it won't be long until it lets you down!

With some cars, the engine starts so quickly that it is difficult to read the voltage while cranking. If your car is one of these, you may be able to delay the engine start by holding the accelerator all the way to the floor while cranking. This is what you would do to start a flooded engine, and it reduces fuel delivery. Since the engine is not really flooded, it will not receive adequate fuel to start running, especially if the engine is cold. We must point out that this technique is not successful on all cars, however. Some Asian and European imports will quickly start anyway, even at full throttle!

Figure 6.3—This swollen battery case is a sure sign of trouble! It indicates that the battery has frozen or severely overheated internally. Don't even think of jump-starting this car!

If the open circuit voltage shows a low state of charge, the charging system should be tested.

1. Start the engine and read the battery voltage. It should be between 13 and 15 volts.

2. Now, let the engine warm up and come to a normal idle speed. The voltage should remain normal, between 13 and 15 volts.

3. Turn on the headlights and air conditioning. The voltage should still remain normal. If so, the charging system is functioning normally. If not, a problem in the alternator is indicated. Although it is possible for an alternator to show normal voltage and still not

JUMP-START YOUR OWN CAR?

Hooking up a set of jumper cables seems simple enough, and it is. However, it is one of the easiest ways to damage the myriad of electronic components found on late model vehicles. If the cables are inadvertently reversed, even for a split second, anything from the alternator to the instrument cluster can be damaged beyond repair. Repairs from this type of damage commonly run anywhere from $600 to $2,500, depending on the extent of the damage and the cost of replacement parts. Jump-starting is also risky from the standpoint of personal injury. Even today's maintenance-free batteries produce explosive hydrogen gas that can be ignited by a single spark, and cause the entire battery to explode violently. Unless it is a true emergency, we recommend leaving jump-starts to the road service professionals. If you must jump-start your own car, remember that you have only one chance to get it right! Adhere carefully to the following guidelines:

- Protect yourself! Safety glasses are an absolute must, and gloves are highly recommended to avoid contact with battery acid and jumper cables that sometimes get hot.

- Use only high quality, heavy-duty jumper cables. The wires in the cables must be at least four gauge to carry full starting current. (The lower the gauge number, the larger the wire.) Otherwise, you will be depending on the dead battery's ability to accept a surface charge in order to crank the engine. Cheap cables use smaller wire, which overheats easily and may actually melt.

- Make absolutely sure that everything in the car is turned off, including the ignition. Unplug ancillary devices, such as cell phones. Our best advice is to have the keys in your pocket while connecting the cables.

- Be absolutely sure that you can identify the positive and negative battery terminals on both cars. If in doubt, check to see which is which with your multimeter.

- Connect the cables one at a time. First, connect the positive cable to the battery positive terminal on the donor vehicle, then to the positive terminal on the dead vehicle. Then, connect the negative cable to the battery negative terminal on the donor vehicle. **Finally,**

connect the negative cable to the dead vehicle, but not to the battery. Instead, connect to an unpainted metal surface on the engine that is at least two feet away from the battery, if possible. If you can, position yourself so that you are facing away from the dead vehicle's battery when you make this final connection.

- Wait at least one full minute before attempting to crank the dead vehicle's engine. This allows the dead battery to obtain some surface charge and helps lessen the electrical load on the jumper cables.

- Once the engine starts, allow the vehicle to run for a few minutes with the jumper cables still connected. If the battery is completely dead or has an internal problem, the engine may not continue to run when the cables are removed. Believe it or not, the alternator needs a small amount of power from the battery, or it cannot charge. Many late models cannot run without at least a partially functioning battery in the vehicle. If you encounter this situation, you must either install a new battery wherever the car is, or have it towed.

- Disconnect the jumper cables by reversing the order in which you connected them. Remove the negative cable from the dead vehicle first, and the positive cable from the donor vehicle last.

charge properly, it is very unlikely. The charging voltage test is so reliable that most manufacturers now install a voltmeter rather than an ammeter in the instrument cluster. Unfortunately, the voltage scales used on these gauges are usually vague, at best. Often, it is difficult to tell just what is a normal reading, and it is impossible to read specific voltage numbers.

Current Draw

In the event that you see a low state of charge with a normal charging system, it is possible that something in the car is drawing current from the battery abnormally, or, more likely, that something is still turned on while you are testing. If you are absolutely certain that everything is turned off, it is best to seek the aid of a professional technician to pinpoint the cause of a current draw. You should also note that a single symptom is consistently present when an excessive current draw exists:

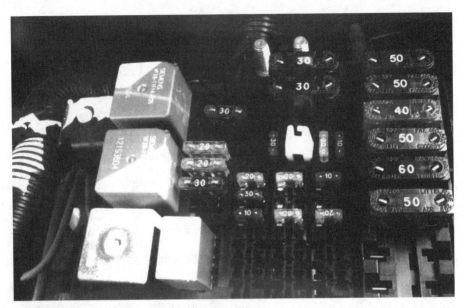

Figure 6.4A—The ATC is the most common type of fuse used on today's cars and trucks. It can be tested without removing it from the fuse block—a handy little feature!

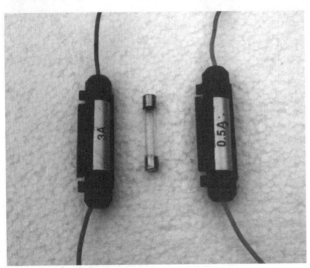

Figure 6.4B—AGC fuse. Note that the two fuse holders have different ratings, although they are the same physical size.

the engine will crank slowly or not at all after the car has been sitting unused for a while. Generally, the longer the car sits, the worse the condition. If you experience the need for frequent jump-starts, even with a fairly new battery, suspect a current draw.

CHOOSE THE RIGHT FUSE

When you encounter a blown fuse that must be replaced, it is absolutely essential that you replace it with the correct type and amp rating. Some "slow-blow" fuses are designed to tolerate slight, momentary overloads without blowing. Most are not. This feature is important, because some circuits normally produce a surge of electrical current as they turn on or off. This surge will cause a standard fuse to blow, even though nothing is really wrong. On the other hand, installing a slow-blow fuse where it is not needed may not provide adequate protection for today's tiny circuit chips.

It is also critical to use the correct amp rating, stamped on the fuse itself and listed on the fuse holder or in the owner's manual. Make sure that the number on the fuse being installed is not higher than the listed rating. Otherwise, you will be allowing more current to flow through the circuit than it was designed for—a recipe for electronic disaster that will likely result in expensive damage!

Bad Connection, Internal Problem, or Relay Failure

What about an engine that will not crank at all, yet the battery voltage indicates a full charge? In this case, it is possible that the battery is, indeed, fully charged, but not capable of operating the starter because of a bad connection or internal problem. It is also possible that the starter motor or relay has failed.

As a quick check, turn on the headlights and attempt to crank the engine. If the headlights go out as soon as you turn the key to the start position, you have a defective battery or a poor connection at the battery terminals. If the terminals appear dirty or corroded, you may be able to restore normal cranking by cleaning them. Time to get out the keep-alive unit, the battery cleaning brush, and the safety glasses. (You should already be wearing them if you are under the hood at all!)

Install the keep-alive as we discussed earlier, and loosen the negative battery terminal clamp nut or bolt using a properly fitting wrench. Then, gently remove the clamp from the terminal. If necessary, spread the ends of the clamp apart slightly, using a large screwdriver, to release it from the post. You will not have to worry about this with side-mount terminals. As soon as you unscrew the retaining bolt, the cable end terminal will fall off!

Remove the positive terminal in the same manner. Then, clean both terminal clamps and the battery terminals themselves using the brush. Heavy corrosion can be removed using a commercial spray-on battery cleaner from the parts store, or a paste made at home from baking soda and water. An old toothbrush works well to apply the paste. It is also a good idea to wipe off the battery case while the terminals are removed, but don't get any cleaner down into the cells! After cleaning, reinstall the terminal clamps, positive first, then negative. Remove the keep-alive and try to crank the engine. If a bad connection was the problem, it will crank normally. If it does not, the next step is a jump-start. A successful jump-start means that there is an internal battery problem. A defective starter motor or relay still will not crank the engine, even with jumper cables connected. If you have either of these problems, it's time to call the tow truck!

We advise against jump-starting your car yourself in nearly every case. There are exceptions, and we will explain them, but jump-starting a car is tricky business because you can do a lot of damage if you don't get it right.

Most inexpensive jumper cables are intended for use when an engine will not crank because the lights were left on or the battery is just weak. They are not heavy enough to start a car with a battery that cannot accept a charge. The car-to-car charging devices that plug into the cigarette lighter are considerably safer than jumper cables, but they are also completely useless in this case. Finally, consideration must be given to the potential for damage to any number of electronic components and the danger of a battery explosion if the cables are reversed or the defective battery is shorted. We cannot, in good conscience, recommend jump-starting your own car unless you are fully qualified to do so, or in the case of an emergency.

Engine Cranks But Won't Start

In the case of an engine that cranks normally but will not start, there are a few tests to perform that may help you get running again. The first thing to do is to think about what, if anything, was happening the last time the engine was running.

Inertia Switch

A car that experienced a minor collision, panic stop, skid, or very rough washboard roads just before stalling may have stopped running due to a tripped inertia switch, a safety device that is used to shut off fuel to the engine in case of a collision. Some of these devices are very sensitive and may be triggered by any of the above conditions. Not all cars have inertia switches, but they are found on virtually all Ford, Mercury, and Lincoln cars and light trucks since the early 1980s, on some models of Jaguar, and on a few General Motors light trucks. Your owner's manual will tell if your car has one and where it is located. If the inertia switch has tripped, pressing a reset button is all that is required to get the car started.

DON'T TRUST THE GAUGES TOO FAR!

It can be a costly mistake to assume that the gauge is faulty if an abnormal reading is seen. But gauges can lose their accuracy, or quit working altogether. Situations such as an empty fuel tank with a gauge that reads half full or a normal engine with the oil pressure warning light on happen everyday! Roy says: "Most dashboard gauge units take less than two cents' worth of materials to manufacture. They are designed to be indicators, not to be completely accurate!" Don't ignore them, however!

Blown Fuse

If the inertia switch is okay, or your car doesn't have one, check to see if the fuse that protects the PCM is blown. All cars and trucks have one or more fuse boxes, usually located under the hood, under the dash, or next to the glove box. Your owner's manual will tell you where the fuses are located on your car, as well as which circuit each fuse protects. As a matter of fact, it would be a good idea to look up this information before you have a problem!

Most late models use the ATC type of fuse, which has small exposed test points near each end. To test your fuses, attach your test light clip to an unpainted metal surface somewhere near the fuse block, turn the key to the "run" position, then simply touch the test light probe to each contact of the fuse you wish to test. The test light should light, upon contact, on both sides; a fuse that is blown will illuminate the test light on one side only. If the light does not come on at all, try testing other fuses. If there is no light from any fuse, it is likely that the test lead clip is not grounded properly or the test light is burned out. Try another unpainted metal surface, or take the test light to the battery and verify that it works. With a little perseverance, you will be able to find a good spot to ground the test light lead. By the way, any fuse on the car may be tested in the same manner. Simply turn on the circuit you want to test and use the test light.

If you should encounter a blown fuse, be sure to replace it with the same type and amp rating. Also, you will need to have a technician determine what caused the fuse to blow, or this could be a frequent occurrence! Don't be surprised if he or she goes right to the problem. Remember that the same quirks often exist on many vehicles of the same model and make. The manufacturer may have even issued a technical service bulletin containing a fix for the condition. (Note that this is not the same as a recall. We will deal with the difference in the next chapter!)

Fuel Pump

If the engine still will not start (you have ascertained that there is fuel in the tank, haven't you?) but the circuits appear to be getting power, there is one more test you can perform to try to determine the problem. This is where your vacuum hose comes in. Virtually all fuel-injected engines use an electric fuel pump mounted in or near the fuel tank. To determine whether or not the pump is running, insert one end of the vacuum hose into the fuel filler neck. Just an inch or two past the metal flap in the neck is sufficient. Then, put the other end to your ear and have an assistant turn the key to the "run" position. Crank the engine. You should hear the gentle whirring of the pump for a second or two when the key is first turned on. Then, the pump should run again while cranking. This is another good test to practice while the car is running normally. That way, you will be familiar with the pump's usual sound.

If the pump does not run at all, the fuel pump relay may have failed or the pump itself may be defective. If you can locate the fuel pump relay on your car, it might be worthwhile to replace it and try cranking once more. These relays are usually very inexpensive at the local parts store, and are more likely to fail than the pump itself. Unfortunately, you will probably not find the location of the relay listed in the owner's manual. If your car has a black box under the hood labeled "electrical center" or something similar, the fuel pump relay is likely to be right under the cover, along with several relays. There is usually an index under the cover that tells which relay is which.

If all of the above procedures have failed to get your engine started, it is unlikely that you will be able to get it running without more in-depth diagnostic techniques and the correct replacement part. Special test equipment such as a lab scope and scan tool, designed to perform pin-point diagnosis and read the PCM's data stream, will likely be required. At this level, the job entails hefty investment and a high degree of specialized training, so it's time to call for professional help.

Engine Starts, But Won't Run Properly

Although there are a multitude of problems that can cause this symptom, the two most probable causes for this condition are a sensor that is sending a bad signal to the PCM and insufficient fuel supply to the engine. Restoring normal operation is often as simple as plugging in a loose connector or re-installing a hose that has come loose.

Loose Connector

In the case of an incorrect signal from a sensor, the MIL should be on to tell the driver that something is wrong electronically. If so, check to see if there is a wire connector plug that has come loose from the air filter housing or near the oil filter. These are two common places where things get knocked loose during routine maintenance, and it could take a few days for the plug to actually fall off.

Disconnected Engine Intake Manifold Hose

If the MIL is not lit, it is possible that there is no fuel in the tank (even if the fuel gauge says there is) or that a hose leading to the engine intake manifold has broken or popped loose. This causes air to rush into the intake and makes the fuel mixture much too lean. To check for an air leak, raise the hood and listen carefully to the sound the engine makes as it starts. If an intake air leak is present, you will hear a hissing noise as the engine starts. If so, look carefully for a hose that is broken, or a hole or fitting that has nothing plugged into it.

Further inspection will probably reveal a disconnected hose or plastic pipe nearby. If the engine will continue to run on its own, you may be able to isolate the leak's location using your vacuum hose. Put one end to your ear, as you did to hear your fuel pump run, and move the other end slowly over the engine. Pay particular attention to the throttle body, the metal part where the large air intake hose connects to the engine. You should be able to hear any leaks large enough to cause problems. If nothing is revealed, it's best to summon a technician for further diagnosis.

Overheating

The engine's cooling system is one of the simplest systems on the whole car, but it is often among the most misunderstood. Depending on the year and model, the PCM may operate an electric radiator cooling fan, and perhaps one or two relays may be found in the fan's circuit. Other than that, the cooling system is quite basic and electronics-free. Years ago, virtually all radiator cooling fans were driven directly by the engine using a drive belt, which often drove the water pump at the same time. In fact, the drive belts were commonly called the "fan belts" if you go back far enough in automotive history to pre-date air conditioning and power steering!

Nowadays, there are only a handful of cars that use engine-driven fans, because this method robs the engine of horsepower and decreases fuel economy. It also makes it very difficult to place the engine sideways, which is best for front-wheel drive power trains. Instead, nearly all late models use one or two electric fan motors to move air through the radiator and the air conditioning condenser.

Sometimes, it is difficult to tell when the engine is running too hot. The normal operating temperature can vary quite a bit with different driving demands and varying weather conditions. However, boiling noises and steaming coolant leaking onto the ground are sure signs of trouble. Overheating can be caused by only three conditions: low coolant, lack of coolant circulation, or lack of airflow through the radiator. Let's look at these causes and how to check the system.

Low Coolant

For a quick check of the cooling system, begin by opening the radiator cap when the engine is cold. The system should be full of coolant, right to the top of the radiator. If the coolant is low, it must be topped off using equal parts of the right type of antifreeze and clean water. Also, add the antifreeze and water mix to the coolant surge tank. On most cars, the tank has a cap marked "engine coolant only." If the tank has indicator lines, fill only to the "cold" mark. If there are no lines, fill the tank about half full. You should make checking the coolant level in this tank part of your weekly check under the hood. A system that regularly requires coolant has a leak somewhere, either external or internal, which must be diagnosed and corrected.

Checking the Fan

Next, replace the cap and start the engine. Lack of airflow is actually the most common cause of overheating. Overheating from lack of airflow is usually related to a fan that is not operating. If the car has air conditioning, turn it on and observe the radiator

EXTENDED LIFE COOLANT TROUBLES

An interesting little quirk found in extended-life coolant came to light some two years after General Motors began using it in their new cars. It seems that Dex-Cool does not do well when exposed to air in the cooling system, and causes heavy, rust-like deposits to form in the radiator and heater core if the vehicle is operated for extended periods with a low coolant level. GM has issued a service bulletin that addresses this issue, but there is no recall. One more reason to keep that cooling system topped off! Remember to check it weekly!

cooling fan(s). On nearly all models, at least one radiator fan will start when the air conditioner is running. This is a good indicator that the fan is at least capable of running. Now turn off the air conditioner and let the engine idle. At least one fan should begin to run after five minutes or so, and before the temperature gauge reaches the high end of the scale. The fan will cycle off and on every one to two minutes on a normal system.

If the fan does not run, check out the fan circuit. First use your test light and see if the fan is getting battery power. Most fan motors are controlled by switching the negative, or "ground side," of the circuit off and on, while battery voltage will be present to the fan whenever the key is in the "run" position. With the key turned off and the engine cool, locate the wires coming from the fan motor and trace them to the nearest connector. Keep your hands clear of the fan blades. Unplug the connector, being careful not to damage the retainer clips.

Now clip your test light lead to the battery negative terminal or an unpainted metal surface on the engine. Touch the probe to battery positive to make sure the light works. Turn the key to the "run" position, then go to the connector you just unplugged. Touch the light probe to each terminal in the half of the connector that *does not* lead to the fan. One of the terminals should light the test light. If so, you have power to the fan. If not, there may be a blown fuse or a basic wiring problem. If your car has more than one fan, test for power to the other motor in the same manner.

If the test light comes on, get your multimeter and set it to read ohms; set the range to 20,000 (or the closest range setting your meter has to 20,000 ohms). Now hook the meter leads to the terminals on the half of the connector that *does* connect to the fan and read the meter. If the fan is defective, your meter will read infinite ohms, an open circuit. On most multimeters, this is indicated by a flashing display, and perhaps a reading such as *OL*, "over limit." In this case, the fan motor must be replaced. A reading of between 4 and 20 ohms indicates that the motor is okay. The problem is in the fan control circuit or the PCM. Like the electric fuel pump we mentioned earlier, the fan control circuit usually contains a relay, although a few models use electronic control modules. The relay is generally found under the hood in the electrical center. It is an inexpensive part to replace. If that doesn't do the trick, it's time to make an appointment at the repair shop.

Coolant Circulation

If the fan runs, but the engine continues to get hotter and hotter, the coolant is not circulating. Suspect a defective thermostat or water pump. Since many engines use a drive belt to run the water pump, a loose or broken belt could be the problem. Take a careful look. If the belt is okay, the radiator could be clogged, or it may be necessary to remove the thermostat or water pump for further testing. You will probably want to refer the job to a repair facility at this stage.

Other Coolant Considerations

It's important to use the correct type of antifreeze, in the right mix with water, and to change it when necessary. How often the coolant needs to be changed depends on a number of factors, including the cooling system design, the specific blend of additives in the coolant, and whether or not it is one of the newer extended-life formulas.

The vast majority of cars on the road today still use ethylene glycol–based coolant, which has been the industry standard for many years. The old rule of thumb was that this coolant should be changed every two years or 30,000 miles, whichever came first, although there were exceptions. Many newer cars use extended-life coolant such as Dex-Cool, which can remain in the cooling system for up to 150,000 miles, but if your car is one of these, proceed with caution. The keywords are "up to 150,000 miles."

Read your recommended maintenance schedule very carefully to find out which coolant to use and how often to change it. It is a good idea to stay with the originally specified coolant for your car, and to maintain a fifty-fifty mix of coolant and water. If your engine will be exposed to extremely cold temperatures of more than forty degrees below zero Fahrenheit, you will have to use a stronger mix or change to a different type of antifreeze. Seek advice from your local dealer or parts store.

Checking Your Coolant

Here are a couple of methods to determine if your coolant's additives are still good, but these tests are not intended to alter your maintenance schedule recommendations! First, you can use a test strip, which works rather like litmus paper. These small strips are available in most auto parts stores. Buy the smallest pack you can; they do not have a long shelf life. You simply dip the strip into your coolant, then shake the excess off and immediately compare the color to the chart provided. The readings indicate the freeze protection and the "reserve alkalinity" of the coolant. The higher the alkalinity, the better the condition of the additives.

The second test sounds a bit crazy. It is not widely embraced by the auto service industry and you won't find it in the textbooks, but it works. This time, you will be using your multimeter to see if the coolant in your engine is allowing any electrical activity in the system. That's right: believe it or not, as the additives are used up, the coolant becomes more acidic and a small voltage is produced as the metal components begin to corrode. It's sort of like a very low-grade battery! To see what is happening, set your meter to volts and select a low range, such as zero to one volt. Connect your negative test lead to the negative battery terminal.

Now remove the radiator cap (with a cold engine, of course!) and dip the tip of the positive test lead right into the coolant. The closer to a reading of zero, the better. The higher the reading, the more corrosion is happening inside your cooling system. The danger point is about 0.5 volts for most cars. If you read 0.4 volts or less, your coolant should be safe for the time being. It does not matter whether the minus sign appears next to the reading or not. You are only measuring the amount of voltage. If you leave the meter connected for a few minutes, you will likely see the voltage start to change. Use the reading you obtained when you first inserted the positive test lead.

Changing Coolant

Changing your own coolant is something that you can still tackle yourself, but we do not recommend it for two reasons. First, it is difficult to get all of the coolant out, and it is equally hard to remove all of the air when refilling the system on many newer vehicles. Second, disposing of the old coolant is a serious concern. Antifreeze is toxic, and it is very dangerous if it leaches into the soil or water supply. Even the coolants advertised as "environmentally friendly" are still toxic, just not as poisonous as the older types! For about the same price as a transmission "total fluid change" service, you can have the cooling system drained, flushed, and refilled by a machine that connects directly to your radia-

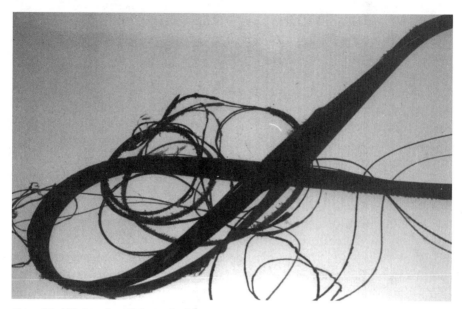

Figure 6.5—This frayed, tangled mess is all that remains of a drive belt that was not replaced soon enough! When a belt squeals, or appears cracked or brittle, it's best to check it without delay.

tor. You can then rest assured that the coolant is automatically being recycled. In some service systems, the same coolant is actually filtered and restored with new additives—recycling on the spot!

Poor Driveability

Perhaps we should clear up the meaning of this word. "Driveability" is a term coined by auto makers in the early days of emission controls to replace "performance." Back then, engines were being de-tuned to make them run cleaner at the price of reduced horsepower and rougher idle quality. "Driveabiltiy" simply means an engine that is functioning as designed, with smooth power application, no surging or stumbling, and an acceptably smooth idle. Nowadays, "performance" has once again become a popular term, as the horsepower and totally smooth operation are back.

In cases of poor driveability, the first thing to do is to carefully observe just what is different than usual. Pay special attention to exactly what driving conditions, if any, trigger the objectionable condition. A cold engine? Warm engine? Accelerating? It makes a difference, because different malfunctions will cause different symptoms. Also, take note of whether the MIL comes on and, again, under what driving conditions. If the MIL is lit, even intermittently, then it is certain that the PCM has seen something abnormal and stored a trouble code. This usually, but not always, indicates a problem in the electronic sensors or controls, and gives the technician a rather specific troubleshooting procedure to follow. If the MIL does not light, the problem is more likely to be in the basic engine systems. The technician is required to perform "no-code diagnosis," essentially troubleshooting the basic engine systems rather than delving into the electronics.

We do not recommend trying to retrieve your own trouble codes or doing your own electronic troubleshooting. If your car is a 1996 or later model, you will almost certainly need special equipment, such as a scan tool, to read the codes in the first place. Even if you can get the codes and a list of what they mean, there is much risk involved in replacing components based solely on the trouble codes, at least from a financial standpoint. There is usually more than one possible cause for each code, and parts stores do not like to allow returns on electronic parts if the attempted repair is unsuccessful. Since one malfunction can sometimes trigger several codes, you could waste a lot of time and money replacing parts that you don't need to replace.

When to Call for Help

The decision to call for help in a given situation is always a judgment call, no matter who is involved. Even professional technicians are relying more and more on help from their colleagues in the form of electronic information services and Internet Web sites that allow them to post questions and answers about diagnosis and repair procedures. With nearly three quarters of a million pages of service information flying around out there, it's no wonder. No one can read fast enough to keep up with everything, and no one has enough room for a library of that size!

Sometimes, the call for help makes sense simply because your time is valuable and it is the quickest solution. Without special training, tools, and equipment, it will take substantially longer to troubleshoot and correct even a basic problem. Of course, there are lots of other reasons, too. Your personal safety and risk of damage to your car are of the utmost importance. You should never attempt anything that you feel uncertain or uneasy about doing. It could be quite dangerous to do something as simple as change a tire or put fuel in the tank, depending on where the car is located.

Change Your Own Flat Tire?

If you are in reasonably good physical condition and are comfortable taking basic things apart and putting them together again, you can most likely change your own tire in an emergency. The key to success lies in knowing how to use the tire changing equipment that is supplied with your car before you are caught off guard on a dark road with an unexpected flat. If you inspect and maintain your tires regularly, you will be less likely to have a tire failure, but trust us, it can happen to anyone! All the care and maintenance you can provide will not stop a loose spike on a railroad crossing from taking out your new premium tire! In case you may be the victim of such an unpleasant surprise someday, it is an excellent idea to practice using the jack and changing tires right in your own driveway.

WHERE TO PUT THE FLAT TIRE?

We were surprised to learn that a few newer European luxury roadsters have an interesting feature (if you can call it a feature). It seems that the spare tire supplied is a space-saver type that stows in the trunk quite conveniently, but the full-sized tires installed on the car will not fit in the trunk, even when flat. Believe it or not, the manufacturers supply a plastic bag to put your flat tire in, and they expect you to carry it with you in the passenger compartment. With a passenger riding along in a two-seat roadster, this could make for an interesting trip, to say the least! This is the type of quirk you will almost never learn from the salesperson unless you ask!

The owner's manual and information found on placards near the jack and spare tire should provide adequate step-by-step instructions on where to place the jack; how to remove the wheel cover; how to raise and lower the car; how to loosen and tighten the lug nuts; and how to

SPARE KEY IN WALLET

You will be amazed to know that, according to AAA, about four million people get locked out of their cars every year. It is one of the most common "help" calls. As soon as you take possession of your car, get a spare key made. Most car keys today are kind of bulky, with big plastic heads, but your licensed locksmith can make you a flat key. Have him or her do it, put that key in your wallet, and forget about it. You may have to throw your wallet in the tray when you are traveling by air, but when you get locked out accidentally you'll be glad you did this—very glad. When you are stuck outside of Mobile, and your spare key is in Memphis, you will certainly have the Memphis blues again. So, just do this, okay?

store the spare tire and jack after everything is back to normal again. However, there are a few things you may not find in the instructions. First, the car needs to be on level, or nearly level, ground that is free of loose sand or gravel. Any other surface invites the car to slip off of the jack.

Second, be sure to loosen all of the wheel lug nuts at least one turn while the tire is still on the ground. This prevents the wheel from turning and stabilizes the jack while you are breaking the nuts loose. After you install the spare tire, tighten the lugs as much as you can while the car is still raised, then lower the car and do the final tightening with the tire on the ground. Since you will not be able to properly tighten the lug nuts to specifications with the wrench supplied, you should have the flat tire repaired and reinstalled as soon as possible.

The days of the full-sized spare tire are over for most of us. Most spares are designed for temporary use only, which means that they should not be driven for more than seventy-five miles or at speeds over fifty miles per hour.

Unfortunately, there are many variations in the design of jacks, wrenches, and even the spare tires themselves—so many that we are unable to give any more than the vaguest of general guidance. Our best advice is to dig into the owner's manual, remove the spare and the jack, and look everything over. If you find it too confusing or feel uncertain in any way as you proceed, don't try it! Just put everything back like it was. There are still ways that you can learn if you want to. Look for a consumer's automotive workshop or basic maintenance course at your local community college or vocational school. Many repair shops and auto clubs provide such seminars periodically as a means of enhancing their image and promoting their business. It's a great sales tool! Wouldn't you tend to trust a repair shop that actually *wants* you to understand how your car works?

What the Tow Truck Driver Can Do

If you must call for road service or a tow, you may feel somewhat helpless or intimidated, but you can still maintain control of your situation. The good news is that tow truck operators are more caring and competent than ever. The vast majority are well-trained professionals who can provide you with damage-free towing, lock-out service, tire changes, and jump-starts. Motor clubs and professional towers' associations work diligently to provide training and certification programs designed to help drivers provide excellent service to their customers.

In many cases, you will need to allow the driver to move your car quickly because of safety and traffic concerns. While it may be all right to put fuel in the tank, it simply is not safe to change a tire on the shoulder of a busy freeway in rush hour (not to mention how this would affect the traffic flow). Davidson has a friend whose brother was killed doing that very thing. Also, for every five minutes that a disabled car stays in the breakdown lane of an Interstate highway, traffic can back up for three miles or more, just from people slowing down to see what is happening. Sometimes, it is best to move the car at least as far as the first exit ramp and then reassess the situation.

The tow truck driver can often perform some quick checks that might get your stalled engine running again. If not, he or she will likely offer you some options on where to take your vehicle. Remember that this is your vehicle, and you will be the one paying for repairs. Ask the driver what your options are and make sure you understand them clearly, including what you will be charged. Ask where the nearest dealer for your vehicle's make is located. If you wish to use a particular repair facility, ask them to tow your car there, but be aware that you may have to pay at least part of the charges, even if you belong to a motor club or your car is under warranty. Most motor club memberships have a maximum number of miles for which they pay towing charges. You pay the rest. Most warranty programs call for towing to "the nearest approved service facility." If you want to go across town, you will have to foot the bill.

In case you are stranded in a strange area, the driver may be able to provide you with assistance in obtaining transportation, lodging, or use of a phone, as well as helping you decide where to have your car fixed.

CHAPTER 7

GETTING IT FIXED!

Getting your car fixed certainly ranks as one of life's more frustrating experiences. The frustration is built in because, first, your car broke down and caused you serious inconvenience; second, it's going to be an unexpected expense; and third, you won't have the use of it while it's in the shop. The days of "loaner" cars are pretty much over and, while most shops try to expedite service, if the shop is busy, it's first come, first served. That can result in the added expense of a car rental, or the further frustration of competing with your spouse for the use of his or her much-needed car. There is almost nothing about having your car in the shop that isn't a pain.

A further potential frustration can arise when trying to explain to the service manager your car's symptoms. If you don't describe your car's symptoms clearly and accurately, your chances of getting it correctly fixed are reduced. Faulty communication between car owner and service technician is the number one cause of incorrect car repairs. Incorrect repairs are, in turn, a major cause of bad blood between car owners and automotive repair shops and technicians.

TECHNICIANS' PAY

Most car owners know that labor and parts go into making up an estimate and final bill, but actually the process can be quite sophisticated. Most technicians are still paid using a system called **flat rate billing,** which was originally designed many years ago to measure the productivity of one mechanic against another.

Today's auto shops employ several flat rate manuals; they are published by Motor Information Systems, Chilton Books, Alldata, or Mitchell. Every repair or maintenance operation on a particular make and model of automobile is allotted a certain amount of time, carefully arrived at by time-motion studies of auto techs at work. Let's say that the replacement of a water pump on a '93 Honda Accord is budgeted at 0.8 hours (48 minutes). That's the amount of time it will take an ordinarily competent technician to do the job. The time includes picking up and putting down tools, some make-ready time, etc.—the whole job from start to finish.

Car manufacturers publish flat-rate manuals, too, but with this important difference: Since their manuals are mainly for warranty repairs that the car manufacturer must pay for, they take a more narrow view of the amount of time involved. They don't count time spent acquiring the part, picking up and putting down tools, opening the hood, putting the car on the lift—things like that. They count only the time that the technician is actually handling the pump and associated parts. They may say that it really only takes 0.6 hours to change that pump—or even less! Needless to say, the only shops that use these manuals are the car dealer's service departments, and even then, usually, only when they are making warranty repairs.

The technician who changes the pump is paid at so much per flat rate hour. The amount depends upon the technician's experience and skill. A beginner is paid less than a pro with high-level skills and years of experience. Whichever one gets the water pump repair job, they both know that they are going to be paid 0.8 of their hourly rate for it, whether it takes them less time or more time to do the job. So, from the technician's point of view, there is a strong incentive to take less time than the budgeted flat rate amount. It is almost like the old-time piece work pay system used by many manufacturers. A skilled technician who routinely takes less than the budgeted time to do a given job will usually do well on payday. We say "usually" because the skilled tech is the one who is going to get the most difficult and demanding jobs. He or she may very well go over the flat rate time now and again. That's the price of success!

However, a technician of ordinary skills who routinely tries to "beat the flat rate clock" may create problems both for himself and for the owner of the cars he works on. There is no flat rate for fixing mistakes. If the car comes back, the technician fixes it on his own time.

Both the technician's flat rate pay rate and the auto shop's overhead, expenses, and profit margin go into establishing a shop's hourly rate, the rate at which they must bill the customer. This carefully arrived at sum can vary quite a bit from shop to shop and from region to region. The car owner's best bet is to find a shop with the most highly skilled technicians and the least amount of visible overhead.

For good reason: from the car owner's point of view a faulty repair is a perfect waste of time and money, a losing proposition in every respect. From the technician's point of view it represents loss of face and financial loss as well, because technicians don't get paid for fixing "comebacks," the industry term for vehicles brought in again for the same problem. Repair failure has high emotional consequences for all involved. Harsh words, bitter feelings, even blows, can and do result.

Was it always like this? Well, it certainly hasn't changed for the better. Since the end of WWII, cars—along with the rest of our culture—have become more complex with each passing year. Automatic transmissions, power steering, power brakes, disc brakes, fuel injection, and air conditioning all made cars more comfortable to use, over the years, but also served to complicate car repair. Digitalization, while making cars more efficient, has certainly added to the complexity. We've duly noted that

How Technicians Are Trained and Certified

Many car manufacturers now offer apprenticeship programs for entry-level technicians. These programs combine a full two-year training program from an accredited school, such as a community college, with periods of hands-on mentoring at local dealership service departments. Roy used to be an instructor in such a program at GM, which is called the Automotive Service Education Program (ASEP). There are also technical schools like Lincoln Tech (Roy used to teach there, too!), and even full four-year university degree programs such as the one at Northwestern University.

Obviously, a university degree program can cost a lot, while the apprenticeship programs provide a stipend and the cost may be at least partially subsidized by the community college system of the state. Tuition can be as low as $35 per credit hour or even less! The cost of education really runs the gamut.

It is not an easy course of training. Today's technician must know hydraulics (plumbing, for brakes, automatic transmission, and cooling system) and electronically controlled hydraulics (antilock brakes), electrical, electronic and mechanical systems (engine, transmission, drive axle, suspension), trim and upholstery, refrigeration, audio systems, tire construction, function and failure, precision machinists' measurement techniques, and on and on.

The main certification body for automotive technicians in America is the National Institute for Automotive Service Excellence, a nonprofit organization that administers over forty different examinations for nearly every known automotive and truck service specialty. About one out of every three applicants fails the certification test. In addition to taking the exams, the technician must demonstrate that he or she has completed two years of work in the service field before he or she can become certified. Not only that, but certification is only good for five years, after which the technician must be retested. ASE certification is a big plus on a technician's résumé.

Once a technician is employed, any number of individual training courses can be taken, through the car manufacturers and/or after-market tool and parts suppliers. Self-study programs are also available, through such private training suppliers as Aspire.

today's cars differ in very important ways from the cars of even ten years ago. Here's a little history and, we hope, some insight into the conditions faced by today's automotive technicians.

A major change was in 1975, the year the catalytic converter was introduced. The catalytic converter precipitated the need for a "feedback" carburetor system to control the fuel mixture—and resulting exhaust emissions—more precisely. Chrysler responded with the computer-chip controlled "lean burn" system. It was primitive by today's standards, but it was a beginning. By 1982 a host of increasingly sophisticated computer systems were being incorporated into cars: systems that could compensate for malfunction, adjust controls for differing driving conditions and, with appropriate monitoring, help technicians diagnose trouble spots.

Not much is known about this '84 Chevrolet S-10, but it appears to be "just broken in" at a little over 100,000 miles. The new owner just repainted it and rebuilt the engine. A great little truck with no monthly payments!

MANHOLE COVER

Roy tells the following story. It seems so silly it must be true:
We had a customer who kept bringing her car in for a thumping
noise when making a left turn. She always dropped the car off after
closing, to be worked on the next day. Two other technicians and I
went over the car from bumper to bumper at least four times, looking for anything loose that could shift or "clunk" as the car was
turning. The inspections revealed nothing and we were never able
to duplicate the noise on a road test.

Finally, I phoned her and asked if she would take me for a ride to
see if she could point out the noise. At this point the plot began to
thicken, because she told me the noise occurred every morning as
she made the first turn from her street onto the main road leading
out of her neighborhood. That was the only time she heard the
noise! She explained that this was the reason she always left the
car at night, so that we could hear the noise first thing in the morning!

I showed up at her house the next morning for the test-drive. She
started the car, backed out of the driveway, and started down the
street. The car purred like a kitten. Not even wind noise. We
stopped at the stop sign, and then negotiated the first left turn.
Sure enough, a loud "thump, thump" was heard beneath the car—
caused by the manhole cover we had just driven over! Struggling
to keep a straight face, I gently explained the situation to her. She
remained a loyal customer for life!

This computerization called for a new type of technician. In the old
days, most mechanics were either self-educated, or educated by older,
more experienced mechanics. Now the field was so complex that they
had to go to school. Automotive service required a type of people who
could be students, who could learn from books and lectures, and apply
themselves to a course of study. The ability to think with one's hands

was not enough any more, although having that ability couldn't hurt if one had all the other qualifications.

One result of all this training was to isolate the technician in the world of his or her own special vocabulary. That's fine when he or she only has to talk to other auto technicians. It's not much help, however, when the technician must talk, and—even more important—listen, to the car owner who is, after all, the technician's first major source of information about a given automotive problem.

Repairs can go sour for reasons other than faulty communication, for sure. Intermittent problems—among the hardest to fix—can daunt the best technicians, and there are cases where a problem is just incorrectly diagnosed. But most bad repairs are the result of some type of breakdown in communications. We are here to help.

Before You Take It In...

We recommend going back and reviewing chapter four, "Getting to Know Your Car." Take particular notice of the section on "Learning the Lingo." This is important. You will have a much better chance of an effective repair if you can describe the problem in terms that the service technician readily understands.

Also, the simple steps that follow can greatly increase your chances for a successful repair experience. To help the technician understand the problem completely, gather as much information as you can before you bring the car in for service. When analyzing the problem, ask yourself the following questions:

- When does the problem happen? For example, is it right after the car is started, or after it is fully warmed up?

- What types of driving conditions cause it to happen? Is it related to acceleration? Stopping? Bumpy road? Turning left or right? Shifting gears?

- In the case of a noise or vibration, exactly where does it seem to be coming from? Be as specific as possible, and try to describe the problem clearly.

- When you think you have a pretty good idea of what the problem is, write it down!

As we have explained, your car contains numerous systems and components: electronic, hydraulic, mechanical, chemical, and structural elements. It is easy to see that a vague description ("There's a noise under the car," "It just isn't riding right...") can open the way for the technicians to misdiagnose and repair the wrong problem, or can result in the conclusion of "unable to duplicate complaint."

As you now know, today's high-tech power train management systems have made great strides in self-diagnosis; in many cases a quick electronic test can point straight to the cause of a problem, even an intermittent one. However, there are still many instances where the diagnosis must be done based on your description of the problem and the technician's knowledge of your car's many systems and parts.

Taking It In

Always try to take the car in during regular business hours, when you can talk directly to the service advisor. Always get the name of the service advisor and, in general, restrict your initial discussions to him or her. It is very frustrating to have to explain your problem to one person after another. Avoid the "early bird" drop-off system. If you must do it that way, take the car in a day or two beforehand and discuss the problem with a service advisor. Write his or her name on the key-drop envelope to make sure that he or she will handle your repair personally.

When you talk to the service advisor, try to give him or her as much information as possible, but don't offer specific repairs that you think will fix the problem. For example, don't begin describing the symptoms by saying "It must need a wheel alignment because..." or, "I think it needs a tune-up." The service advisor or technician may zero in on phrases like this and miss the actual symptoms you are trying to describe. Or he or she may try to sell you maintenance service that has nothing to do with the problem you are reporting.

During your discussion with the service advisor, be sure to ask if the manufacturer has issued a technical service bulletin for the symptom your car is experiencing. If so, the manufacturer may have extended the warranty on the repair that is needed, or may at least share the cost of the repair. To take advantage of such discounts, you will have to have the repair performed by an authorized service department, usually at a dealership.

If a manufacturer's recall has been issued for the problem you are having, it means that a large number of vehicles have been afflicted with the same problem and that it is potentially a serious safety or reliability issue. Most recalls are instigated by a large number of complaints to the Department of Transportation and are usually well publicized. You should receive a written notice of any recall that applies to your vehicle and there should be no charge for the dealer to make the necessary corrections.

Beware of suggestions from the advisor such as, "Let's do a tune-up and see if that fixes it," or ploys such as, "Your car definitely needs a wheel alignment. We believe that may fix the pull to one side." That could mean that he or she doesn't really know what is wrong and may not know how to further diagnose your car if what he or she recommends initially doesn't fix it. It could also indicate that they are trying to sell you a service even if it has nothing to do with your problem. Roy refers to this method of diagnosis as "throwing parts at the problem."

Make sure you see the repair order. And make sure your description of the problem—the way you understand it—is written on it. If there isn't adequate room on the repair order, write a short note and attach it. Remember, the description of the problem that you wrote when you were analyzing it? Bring that with you when you take the car in.

THE FIVE O'CLOCK SURPRISE

There was a grinding noise every time you made a right turn. The service manager nodded sagely when you told him about it and said, "Leave it to me. You can pick it up tonight before 5 P.M." So you get somebody at work to give you a ride to the shop to pick up your old Dodge Caravan Friday at 5 P.M. The service manager hands you a bill and says, "See the cashier. I'll have your car brought around front."

You look at the bill: CV joint, parts, and labor come to just under $400. Yikes! You shove a well-used VISA card over the counter, hoping that there's enough credit left to meet the bill. Out the window you can see somebody pulling up in Old Bessie. You sign the slip, head out into the driveway, open the door, remove that protective paper that the technicians use to keep the seat clean, and climb in. You pull out of the yard. Home is a straight shot left, ten miles down the road. Then you make a right turn onto your street.

What's that? It's the same sound! You make another right turn. Aaaaargh! There it is again! Wait! You just spent $400 to get this thing fixed! They put in a new CV joint! They said that would fix it!

Welcome to the five o'clock surprise! You run to the phone and call the repair shop. A nice voice informs you that the service department hours are Monday through Friday, from 8 A.M. to 5 P.M. Have a nice weekend!

How different the ending of this scenario might have been if only the service manager had taken a short road test to verify the complaint, and another to make sure it was fixed after the repair. By insisting that the problem was verified and clearly understood, you could have avoided the five o'clock surprise, the service manager could have avoided the image of incompetence, and the technician could have avoided what is sure to be a Monday morning comeback—with attitude!

It could use a coat of paint, but only very minor cosmetic surgery is needed on this nice 84 Nissan 300ZX. At 146,000 miles, it is undergoing restoration by its new owners.

Offer to take a ride with the service advisor or technician if the problem is hard to describe or may be difficult to duplicate. Some repair facilities have a road tester or quality control specialist whose job is to identify problems for the technician and then make sure they are corrected when the job is finished. Do whatever you can to assure that the technician gets as much to work from as possible.

When the repair order is written, ask for a written estimate. Most states require that the actual repair cost cannot exceed the estimate by more than 10 percent. Also, ask to have the old parts returned to you, even if you won't know what you are looking at.

Yeah, we know. Who wants this greasy, dirty, heavy junk anyway! But there is a reason for this. If the repair is unsuccessful, or you have any reason to suspect dishonesty after the repairs are completed, showing the old parts to another technician, or having them available as evidence (if it should come to that!) may help resolve the situation. So get a box, or spread some old newspapers down in the trunk, and take the old parts home, for a while, anyway. If they ask you why you want them, just tell them, "My kid loves this stuff."

Of course, you don't want any hazardous materials like a dead battery or four quarts of dirty oil returned to you. No, indeed!

Does the Price Seem Too High?

If you feel that you are quoted a price that is unreasonably high, or that the recommended repair seems unrelated to the problem you reported, just remember that you are still in control of the situation. It is, after all, your vehicle and your money. You only pay for the work you have authorized!

You have several options to consider before giving the okay. To begin with, ask how they arrived at the quoted price for parts and labor. They may have written the estimate based on a standard guide such as the flat rate manual. Or they may be using "menu pricing," which might be based on their previous experience with that make and model, or might be based on nothing in particular!

Also, be sure to ask what kind of parts are being quoted: new, used or remanufactured, and what brand. Remember to ask about the guarantee that is being offered, too. Sometimes the guarantee on parts, labor, or both varies considerably, and that can account for differences in the estimate from one shop to another. If you don't think the estimate is fair, take the car elsewhere for a second opinion. By choosing this option, you may have to pay a diagnostic charge, but you could still save money in the long run.

Once you have a written estimate in hand, you can make a few phone calls to compare pricing and ease your mind. Call some other repair shops and ask for approximate quotes to do the same job, or at least ask how much labor they would charge for it. To compare parts prices shop around by calling several suppliers, including the dealer's parts department, and asking for the list price of the parts on the estimate. Although it is not fair to expect a discounted price, the parts prices on your estimate should not exceed the dealer's list price.

Remember, too, that, despite Benjamin Franklin's famous quote, the cheapest is not always the best. A part obtained from your local dealer's parts department will definitely meet the manufacturer's original equipment specifications, while a cheaper aftermarket unit may not. There may be a significant difference in reliability or performance that will make the more expensive part a better value in the long run. Adopting the mindset of doing repairs as cheaply as possible with the idea that you will soon be trading or selling this car is usually not a good strategy. If the facility repairing your car is going to guarantee their work, they will probably insist on using high-quality parts. We are on the car's side. Use good parts. After all, you're trying for 300,000 miles.

Did They Do a Good Job?

Try to arrange to pick up your car at a time when the person who wrote your repair order is still there, and the service department is not yet closed for the day. That way, should the problem be unresolved, you can insist that they have someone ride with you, if needed, to confirm the problem; someone can document the fact that the repair was unsuccessful; or someone is there to arrange for a refund, if you feel that you were incorrectly charged.

To sum up, the repair facility you trust to take care of your car's needs is genuinely interested in meeting your needs and having you as a regular customer. You and your car are often their only sources of information. The most important single thing that you can do is to clearly communicate your car's problems and help lead them down the path to correct diagnosis and resolution. Unfortunately, terrible repair experiences do happen, and automotive problems that daunt the best of technicians do exist. In these instances, you may have to go beyond the normal channels of communication and resolution to find satisfaction.

If your car repair, or the explanation you are given, is unsatisfactory, the first thing to do is to speak to the manager. In most instances, the manager is empowered to make decisions that the service advisor cannot, such as granting refunds, altering schedules, or arranging for alternate transportation. Most of the time, a satisfactory solution can be obtained at that level. If the manager is unable or unwilling to meet your needs, you still have some choices to consider in reaching resolution. The path to take depends on what type of service facility you are dealing with, and the age and mileage of your vehicle.

If your car is still covered under the manufacturer's warranty, even if you bought it used, you have some options that may not be available on an earlier model. In either case, knowing your rights is just as important as clear communication. We will discuss unsuccessful repairs for vehicles under warranty first, then look at options for those repairs not covered.

Cars Under Warranty: What Outcome Do You Want?

For repairs made under warranty from the manufacturer, you are almost certainly dealing with the service department of a car dealership. Virtually all dealers are tracking their customer satisfaction carefully these days, both for warranty and non-warranty repairs. They have good reason to do so, because the manufacturer of their make provides incentives and bonuses based on the dealership's customer satisfaction index, or CSI. You may have been asked to fill out a survey regarding your last service department visit, which is sent back to the manufacturer. If you gave the service department bad marks, you may have even heard from the service manager or someone else at the dealership, who asked what they could do to improve their rating in your eyes.

If you really believe that the cause of the problem lies with the service department personnel, rather than a defect in the car, you can try speaking with or writing to the service manager's boss, usually the general manager. There may also be a higher boss, the dealership owner. This tactic can be very successful, provided that you do not let your frustration reach the level where you are being abusive or making threats rather than asking for assistance.

That's important, by the way. A bad car fix adds another level of frustration to an already frustrating experience. So it is not unusual to have an emotional reaction. Now is the time to practice self-control, to be polite, and not let things get personal. Diplomacy and politeness cost nothing, and will pay big dividends. You don't want to get service people mad at you; you want them to fix your car. And you want them to *want* to fix your car.

Suppose you have an oil leak or other defect in the car that they just cannot seem to fix. You have two avenues of approach in seeking resolution beyond the level of the service department. Which one you pursue will depend on what you want to happen. Your possible outcomes are:

- To have the vehicle repaired correctly;
- To force the manufacturer to buy back the vehicle so you can get another car.

The first approach involves seeking assistance from the manufacturer in getting the problem fixed, if the dealer has been unsuccessful. To do this, call the nearest zone office or the car manufacturer's customer assistance hotline. If there is a central, toll-free number to call, it can be found in your owner's manual. Once the manufacturer is involved, they may need to investigate your situation. Sometimes you may be asked to take the vehicle to the same, or a different, dealership. You may also be asked to meet with the local zone representative at a specified date and time.

If the problem still cannot be fixed to your satisfaction, even with the manufacturer involved, the second approach is to seek assistance under the motor vehicle lemon laws in your state. At this point, you are asking the appointed consumer protection authority to hear your case and arbitrate resolution between you and the manufacturer.

The Lemon Law

Under federal law, all manufacturers that sell new vehicles in the United States must have a binding arbitration mechanism in place to carry out the terms of the lemon laws. There is no completely uniform procedure and coverage, because the arbitrators must be certified individually in each state, and the terms of the legislation itself also vary from state to state. However, it is nearly always required that you attempt resolution through arbitration before filing a lawsuit in court.

As with learning how your ABS works and how to change your tire, we highly recommend that you obtain information about the lemon laws in your state before you actually need to be involved with them. Generally speaking, the term of protection under lemon law legislation is the same as most new car warranties: three years or 36,000 miles. However, there are some states that only allow protection for 12,000 miles, while others may allow hearing of cases for up to 50,000 miles. In addition, there may be an agreement between the manufacturer and their designated arbitrators that provides coverage above that required in your state.

It is essential that you learn your rights under the law at the first sign that your car may have an irreparable problem or deficiency. Begin by checking the owner's manual. By law, the manufacturer is required to inform you, in writing, how to proceed and whom to contact to begin pursuit of arbitration. This information is usually found at the back of

Motor Vehicle Defect Notification
(Please print clearly in ink)

Pursuant to the Florida Lemon Law, notice is given to the manufacturer as follows:

☐ **The vehicle has been out of service at least 15 days to repair one or more substantial defects.**

☐ **3 or more repair attempts have been made to repair the same substantial defect or condition.**

Description of continuing defect(s) or condition(s) _____

_____

(NOTE: this is not a complete description; the manufacturer should ascertain all appropriate information.)

I am requesting that you make a final attempt to correct the continuing substantial defect(s) or condition(s).

Vehicle Make _____ Model _____ Year _____

VIN _/_/_/_/_/_/_/_/_/_/_/_/_/_/_/_/_/ Date of Delivery _____

Name and City/State of selling dealer or leasing company (if applicable) _____

_____

Name and City/State of authorized service agent(s) attempting previous repairs: _____

_____

Consumer _____ Home phone(___)_____

Address _____ Work phone (___)_____

_____ Signature _____

_____ Date Mailed _____ (1/98)

Figure 7.1—Lemon Law Compliant Form

the manual, at or near the back cover. You can also obtain federal lemon law information from the U.S. Department of Transportation's Web site at usdot.gov or by calling their consumer complaint hotline at (800) 424-9393.

To obtain assistance under the lemon laws, you must follow the necessary steps, and in the correct order. Generally speaking, you will be required to do the following:

1. Be able to show that the problem constitutes a substantial impairment to the vehicle's use or safety. For example, a rough idle would not be substantial impairment in most cases, but repeated stalling or a condition requiring the vehicle to be towed would be.

2. Make reasonable effort to resolve the problem through the manufacturer's normal service procedures.

3. Allow the manufacturer a specified number of reasonable repair attempts for the same problem, usually three, or show a history of impairment to your use of the vehicle from many different needed repairs. Even though the manufacturer should have records of your repairs, you must also provide documentation of all repairs and related events, and receipts for any costs for which you are seeking reimbursement.

4. File your complaint to the arbitrating agency using the proper procedure and, if required, the proper form. You will generally be asked to specify the outcome you are seeking: mandated specific repair of the problem, a buyback of the vehicle, and/or reimbursement of related costs for rental cars, etc.

5. Depending on the specifics of your case, you may be required to make the vehicle available for inspection by the arbitrators and/or the manufacturer.

Once you have taken all the steps, the arbitrators will hear the case and make a decision. There is a time limit specified for this action, so it will not drag on forever, as court cases so often seem to do. The arbitrators may grant what you request, or they may make a different decision. For instance, they may rule a buyback when you asked for specific repairs if they believe that further repairs will not be successful. When the ruling has been made, it is binding to the manufacturer, but not to you. You may accept their decision, or reject it and elect to take further legal action through the courts.

If the arbitration decision is a buyback, and you accept it, the manufacturer will take possession of your car and refund you the full price you paid, less a specified allowance for the miles you have driven it. Once again, this allowance varies from state to state, but you must be given all of the information before you agree to accept the decision. In some cases, you may be able to negotiate a settlement wherein you will be given a new car at no cost (providing, of course, that you would be willing to try another car of the same make). You do not have to sign the acceptance as soon as the decision is made and settlement offered because there is a period of time allowed for you to consider your options. The actual time allotted to think things over varies from state to state.

Repairs Not Covered by Warranty

For unsuccessful repairs that were not performed under the manufacturer's warranty, you generally cannot seek lemon law assistance. This means that you must go straight from the repair facility to the courts. However, there are still some things you can do to seek satisfaction, or at least financial relief. In this case, the best initial approach is to try to resolve the problem directly with the service advisor and manager. The next steps will depend somewhat on the type of repair facility involved.

If the repairs were done by a shop that is part of a chain, either national or local, you can contact their customer service department or head office. If they are genuinely interested in customer loyalty and good word-of-mouth advertising, they are likely to have procedures established for handling complaints. Most businesses realize that a customer who complains is one who is not yet lost. An opportunity still exists to build a loyal customer (or at least one who will not spread negative information about them) if they take steps to do all that they can to meet the customer's needs.

A customer who simply takes his or her business elsewhere without complaining is almost certainly a lost customer. If you feel that you are not being taken seriously, even by the management, writing a letter to their top executives with copies sent to the Better Business Bureau and your local and state consumer protection agencies should get their attention. Of course, if you are dealing with a small independent facility, you are probably standing in the head office, and the service advisor may also be the CEO.

If you are unable to obtain satisfaction through the channels of management, negative publicity may be the best alternative, short of filing a lawsuit. What we're talking about here is to notify the local TV stations and newspapers of your problem and your inability to get it resolved. Even the smaller stations and publications usually have someone assigned to report consumer affairs, and they are always hungry to turn up something they might be able to somehow sensationalize. Auto repairs are among their favorite subjects! You might even find that they will go to the repair facility themselves and try to help you get restitution. If they succeed, not only does it help you, it makes them look like heroes to their audience. On the other hand, the business owner will come across looking very bad if he or she doesn't do the right thing while dealing with the media—very bad indeed!

Alternate Assistance and "Hidden" Warranties

Well, they are not really "hidden," but there are warranties of which you may not be aware that can be of tremendous help if smoke leaks out of that high-tech audio system that you just installed last month, or if the water pump that you replaced six months ago lays nine quarts of coolant in your driveway tomorrow night. We already advised you to ask about the guarantee on any repair job before you approve the charges. There may also be other warranties, which offer more protection. For example, there may be automatic product warranties available simply by paying with a particular credit card. Sometimes you have to ask for these to be implemented at the time of purchase by calling the credit card company, so it's a good idea to ask your credit card providers what is available and to arm yourself with that information. It's a great benefit for *all* kinds of purchases, in addition to auto parts and repairs! If the road to dispute resolution seems impassable, some credit card companies will withhold payment for you until the dispute is settled.

You may also find assistance through your membership in any number of organizations, such as motor clubs. In the United States, the Automobile Association of America (AAA) provides a warranty for their members on all repairs performed by their network of approved auto repair facilities. The warranty is honored nationwide at any AAA-approved repair facility. In addition, AAA provides binding arbitration for members if a dispute with one of their approved facilities arises.

The business of automotive service has become more complex and sophisticated for the technician, manager, and customer alike. This book is dedicated to the customer in the equation, and is meant to give him or her, if not an advantage, at least a level playing field. A well-informed customer has a better chance of attaining satisfactory products and service. When in negotiation for a better product or service, remember: perseverance furthers.

Now we have to leave you with our hopes that you have enjoyed reading this book as much as we enjoyed writing it. We believe that if you read this book and follow the recommendations that we have given you, you will get much more out of your car in terms of reliability, service, and, yes, pleasure. We guarantee it, in fact.

If you read and retain what we tell you about how today's cars function, you will certainly end up knowing more about cars than most car owners. Service people and others who are also knowledgeable about cars will treat you with considerable respect as they come to understand this, and you and your car will benefit as a result.

So, that's it for now. We expect, however, that we will meet with you again down the road. Somehow, we are sure this isn't our last word on the subject of the 300,000 mile car. We hope to keep this book up-to-date, and to be your friendly and accessible source of automotive information as you travel the years and miles with the car of your choice.

But for now Roy has to change the timing belt on his son's 1990 Dodge Dakota. He also would like to get his boat in the water this year—it's been a while. Speaking of boats, Davidson is about to start prowling the southern New England coastline looking for a used Boston Whaler. He's moving back to Connecticut (where he is from originally) and likes to go to Block Island for lunch every now and then if the weather is nice.

In the meantime, happy highways to you all,

— Roy and Davidson

O

OBDII, 136
Octane rating, 66-68
On-board diagnostics, 100
  second generation, 136
Open circuit voltage, 145-147
Open-end lease, 36
Options, 25-26
Output drivers, 126-127
Outputs, 124
Overdrive, 104
Overheating, 156-161
Oversteer, 98
Owner's manual, 65

P

Pads, 116
Parking, 22
Parking brake, 115, 117
Personal Injury Protection (PIP), 13
Pinging, 67
Poor driveability, 161-162, 174
Power steering fluid, 74
Powertrain, 58-59
Power train management system, 122
  control signals, 123-124
  drive-by-wire systems, 132-133
  five signal types, 127-129, 131-132
  input conditioners, 125
  memory, 125-126
  microporcessor, 126
  output drivers, 126-127
  outputs, 124
  self-diagnostic systems, 134-136
  trouble codes, 130
Premature tire wear, 86
Programmable ROM (PROM), 125
Pulling, 98
Pulse-width modulated signals, 132

Q

Quirks, 92

R

R-134a, 112-113
Rack-and-pinion system, 108-109
Radio code, 77
Radio volume changes, 121
Random access memory (RAM), 126
Read-only memory (ROM), 125
Rear-wheel drive, 106

Refrigerant, 112-113
Relay, 110, 127
Relay failure, 151
Repairing your car, 167-173
  alternate assistance and hidden war-
    ranties, 185
  cars under warranty, 180-181
  did they do a good job?, 179
  is price too high?, 178-179
  lemon law, 181-183
  taking it in, 174-179
  unsuccessful repairs not under warran-
    ty, 184-185
  before you take it in, 173-174
Repair order, 175
Resale value, 32
Residual value, 36
Right car. *See* Choosing the right car
Road crown lead, 109
Rotor, 116

S

Safety glasses, 139, 141
Safety in a crash, 21-22, 26
SALA suspension, 108
Scan tool, 135
Self-diagnostic system, 100, 134-136
Serial-data signals, 132
Service advisor, 175, 177
Service brakes, 115
Severe service, 71
Shimmy, 86
Showroom floor tactics, 48-52
Signals (inputs), 123-124, 127-129,
    131-132
Single wire, common ground electrical
    system, 144
Slow-blow fuse, 151
Soft items, 51
Solenoid, 110
Speedometer, 85
Sports utility vehicles (SUVs), 106
Start, 98
Stated purchase option lease, 36
Status-free car, 23
Steering, 59
Steering systems, 108-109
Sticker price, 57
Stolen cars, 14-15
Struts, 108
Stumble, 100
Summer washer fluid, 75